East European Diasporas, Migration and Cosmopolitanism

Following the collapse of the communist regimes in Eastern Europe and the Soviet Union, there were considerable migration flows, the migrations and subsequent diasporas often having special characteristics given the relative lack of migration in communist times and the climate of increasing nationalism which had the potential of working against multiculturalism. This book explores these migrations and diasporas, and examines the nature of the associated cosmopolitanism. It examines issues of identity encountered by migrants, considers the difficulties faced by minorities in being accepted and integrated, and compares East European migration, diasporas and cosmopolitanism with the situation in other parts of the world.

Ulrike Ziemer is a Lecturer in Sociology at the University of Winchester, UK.

Sean P. Roberts is a Visiting Researcher at the Norwegian Institute of International Affairs (NUPI) in Oslo.

BASEES/Routledge Series on Russian and East European Studies

This series is published on behalf of BASEES (the British Association for Slavonic and East European Studies). The series comprises original, high-quality, research-level work by both new and established scholars on all aspects of Russian, Soviet, post-Soviet and East European Studies in humanities and social science subjects.

1. **Ukraine's Foreign and Security Policy, 1991–2000**
   Roman Wolczuk
2. **Political Parties in the Russian Regions**
   Derek S. Hutcheson
3. **Local Communities and Post-Communist Transformation**
   Edited by Simon Smith
4. **Repression and Resistance in Communist Europe**
   J.C. Sharman
5. **Political Elites and the New Russia**
   Anton Steen
6. **Dostoevsky and the Idea of Russianness**
   Sarah Hudspith

7. **Performing Russia – Folk Revival and Russian Identity**
Laura J. Olson

8. **Russian Transformations**
Edited by Leo McCann

9. **Soviet Music and Society under Lenin and Stalin**
The baton and sickle
Edited by Neil Edmunds

10. **State Building in Ukraine**
The Ukranian parliament, 1990–2003
Sarah Whitmore

11. **Defending Human Rights in Russia**
Sergei Kovalyov, dissident and Human Rights Commissioner, 1969–2003
Emma Gilligan

12. **Small-Town Russia**
Postcommunist livelihoods and identities: A portrait of the intelligentsia in Achit, Bednodemyanovsk and Zubtsov, 1999–2000
Anne White

13. **Russian Society and the Orthodox Church**
Religion in Russia after communism
Zoe Knox

14. **Russian Literary Culture in the Camera Age**
The word as image
Stephen Hutchings

15. **Between Stalin and Hitler**
Class war and race war on the Dvina, 1940–46
Geoffrey Swain

16. **Literature in Post-Communist Russia and Eastern Europe**
The Russian, Czech and Slovak fiction of the Changes 1988–98
Rajendra A. Chitnis

17. **The Legacy of Soviet Dissent**
Dissidents, democratisation and radical nationalism in Russia
Robert Horvath

18. **Russian and Soviet Film Adaptations of Literature, 1900–2001**
Screening the word
Edited by Stephen Hutchings and Anat Vernitski

19. **Russia as a Great Power**
Dimensions of security under Putin
Edited by Jakob Hedenskog, Vilhelm Konnander, Bertil Nygren, Ingmar Oldberg and Christer Pursiainen

20. **Katyn and the Soviet Massacre of 1940**
Truth, justice and memory
George Sanford

21. **Conscience, Dissent and Reform in Soviet Russia**
Philip Boobbyer

22. **The Limits of Russian Democratisation**
    Emergency powers and states of emergency
    Alexander N. Domrin

23. **The Dilemmas of Destalinisation**
    A social and cultural history of reform in the Khrushchev era
    Edited by Polly Jones

24. **News Media and Power in Russia**
    Olessia Koltsova

25. **Post-Soviet Civil Society**
    Democratization in Russia and the Baltic States
    Anders Uhlin

26. **The Collapse of Communist Power in Poland**
    Jacqueline Hayden

27. **Television, Democracy and Elections in Russia**
    Sarah Oates

28. **Russian Constitutionalism**
    Historical and contemporary development
    Andrey N. Medushevsky

29. **Late Stalinist Russia**
    Society between reconstruction and reinvention
    Edited by Juliane Fürst

30. **The Transformation of Urban Space in Post-Soviet Russia**
    Konstantin Axenov, Isolde Brade and Evgenij Bondarchuk

31. **Western Intellectuals and the Soviet Union, 1920–40**
    From Red Square to the Left Bank
    Ludmila Stern

32. **The Germans of the Soviet Union**
    Irina Mukhina

33. **Re-constructing the Post-Soviet Industrial Region**
    The Donbas in transition
    Edited by Adam Swain

34. **Chechnya – Russia's "War on Terror"**
    John Russell

35. **The New Right in the New Europe**
    Czech transformation and right-wing politics, 1989–2006
    Seán Hanley

36. **Democracy and Myth in Russia and Eastern Europe**
    Edited by Alexander Wöll and Harald Wydra

37. **Energy Dependency, Politics and Corruption in the Former Soviet Union**
    Russia's power, oligarchs' profits and Ukraine's missing energy policy, 1995–2006
    Margarita M. Balmaceda

38. **Peopling the Russian Periphery**
    Borderland colonization in Eurasian history
    Edited by Nicholas B. Breyfogle, Abby Schrader and Willard Sunderland

39. Russian Legal Culture Before and After Communism
Criminal justice, politics and the public sphere
Frances Nethercott

40. Political and Social Thought in Post-Communist Russia
Axel Kaehne

41. The Demise of the Soviet Communist Party
Atsushi Ogushi

42. Russian Policy towards China and Japan
The El'tsin and Putin periods
Natasha Kuhrt

43. Soviet Karelia
Politics, planning and terror in Stalin's Russia, 1920–1939
Nick Baron

44. Reinventing Poland
Economic and political transformation and evolving national identity
Edited by Martin Myant and Terry Cox

45. The Russian Revolution in Retreat, 1920–24
Soviet workers and the new communist elite
Simon Pirani

46. Democratisation and Gender in Contemporary Russia
Suvi Salmenniemi

47. Narrating Post/Communism
Colonial discourse and Europe's borderline civilization
Nataša Kovačević

48. Globalization and the State in Central and Eastern Europe
The politics of Foreign Direct Investment
Jan Drahokoupil

49. Local Politics and Democratisation in Russia
Cameron Ross

50. The Emancipation of the Serfs in Russia
Peace arbitrators and the development of civil society
Roxanne Easley

51. Federalism and Local Politics in Russia
Edited by Cameron Ross and Adrian Campbell

52. Transitional Justice in Eastern Europe and the former Soviet Union
Reckoning with the communist past
Edited by Lavinia Stan

53. The Post-Soviet Russian Media
Conflicting signals
Edited by Birgit Beumers, Stephen Hutchings and Natalia Rulyova

54. Minority Rights in Central and Eastern Europe
Edited by Bernd Rechel

55. Television and Culture in Putin's Russia: Remote Control
Stephen Hutchings and Natalia Rulyova

56. The Making of Modern Lithuania
Tomas Balkelis

57. **Soviet State and Society Under Nikita Khrushchev**
Melanie Ilic and Jeremy Smith

58. **Communism, Nationalism and Ethnicity in Poland, 1944–1950**
Michael Fleming

59. **Democratic Elections in Poland, 1991–2007**
Frances Millard

60. **Critical Theory in Russia and the West**
Alastair Renfrew and Galin Tihanov

61. **Promoting Democracy and Human Rights in Russia**
European organization and Russia's socialization
Sinikukka Saari

62. **The Myth of the Russian Intelligentsia**
Old intellectuals in the new Russia
Inna Kochetkova

63. **Russia's Federal Relations**
Putin's reforms and management of the regions
Elena A. Chebankova

64. **Constitutional Bargaining in Russia 1990–93**
Information and uncertainty
Edward Morgan-Jones

65. **Building Big Business in Russia**
The impact of informal corporate governance practices
Yuko Adachi

66. **Russia and Islam**
State, society and radicalism
Roland Dannreuther and Luke March

67. **Celebrity and Glamour in Contemporary Russia**
Shocking chic
Edited by Helena Goscilo and Vlad Strukov

68. **The Socialist Alternative to Bolshevik Russia**
The Socialist Revolutionary Party, 1917–1939
Elizabeth White

69. **Learning to Labour in Post-Soviet Russia**
Vocational youth in transition
Charles Walker

70. **Television and Presidential Power in Putin's Russia**
Tina Burrett

71. **Political Theory and Community Building in Post-Soviet Russia**
Edited by Oleg Kharkhordin and Risto Alapuro

72. **Disease, Health Care and Government in Late Imperial Russia**
Life and death on the Volga, 1823–1914
Charlotte E. Henze

73. **Khrushchev in the Kremlin**
Policy and government in the Soviet Union, 1953–1964
Edited by Melanie Ilic and Jeremy Smith

74. **Citizens in the Making in Post-Soviet States**
Olena Nikolayenko

75. **The Decline of Regionalism in Putin's Russia**
Boundary issues
J. Paul Goode

76. **The Communist Youth League and the Transformation of the Soviet Union, 1917–1932**
Matthias Neumann

77. **Putin's United Russia Party**
S. P. Roberts

78. **The European Union and its Eastern Neighbours**
Towards a more ambitious partnership?
Elena Korosteleva

79. **Russia's Identity in International Relations**
Images, perceptions, misperceptions
Edited by Ray Taras

80. **Putin as Celebrity and Cultural Icon**
Helena Goscilo

81. **Russia – Democracy Versus Modernization**
Edited by Vladislav Inozemtsev and Piotr Dutkiewicz

82. **Putin's Preventive Counter-Revolution**
Post-Soviet authoritarianism and the spectre of velvet revolution
Robert Horvath

83. **The Baltic States – From Soviet Union to European Union**
Richard Mole

84. **The EU-Russia Borderland**
New contexts for regional cooperation
Edited by Heikki Eskelinen, Ilkka Liikanen and James W. Scott

85. **The Economic Sources of Social Order Development in Post-Socialist Eastern Europe**
Richard Connolly

86. **East European Diasporas, Migration and Cosmopolitanism**
Edited by Ulrike Ziemer and Sean P. Roberts

East European Diasporas, Migration and Cosmopolitanism

Edited by Ulrike Ziemer and Sean P. Roberts

LONDON AND NEW YORK

First published 2013
by Routledge
2 Park Square, Milton Park, Abingdon, Oxon OX14 4RN

Simultaneously published in the USA and Canada
by Routledge
711 Third Avenue, New York, NY 10017

Routledge is an imprint of the Taylor & Francis Group, an informa business

British Library Cataloguing in Publication Data
A catalogue record for this book is available from the British Library

Library of Congress Cataloging in Publication Data
A catalogue record for this book has been requested

ISBN: 978-0-415-51702-7 (hbk)
ISBN: 978-0-203-08112-9 (ebk)

Typeset in Times New Roman
by Cenveo Publisher Services

Printed and bound in the United States of America by Publishers Graphics, LLC on sustainably sourced paper.

Contents

Contributors

Tsypylma Darieva holds a PhD from Humboldt University, Berlin, Germany. Her research interests include the anthropology of migration, transnational diaspora and homecomings, cosmopolitan sociability and post-socialist urbanism in Eurasia (Europe and South Caucasus). Currently, she is Associate Professor of Anthropology at the Graduate School of Humanities and Social Sciences, University of Tsukuba, Japan and Associate Member of the Collaborative Research Center at Humboldt University, Berlin. Her publications include the monograph *Russkij Berlin. Migranten und Medien in Berlin und London* (Münster: LIT Verlag, 2004).

Rebecka Lettevall is Associate Professor of Intellectual History. Currently, she is Pro-Vice Chancellor at Södertörn University in Stockholm. Previously, she was Research Director at the Centre for Baltic and East European Studies (CBEES) at Södertörn University 2006–10. Her main research areas are Kant and the political philosophy of German Idealism, the intellectual history of the ideas of peace and cosmopolitanism.

Darya Malyutina is a final year doctoral candidate at the Department of Geography, University College London. Her doctoral research focuses on friendship among Russian-speaking migrants in London. Her main research interests include migration, transnationalism, super-diversity of London, social networks, cosmopolitanism, ethnography and feminist geography.

Dorota Praszałowicz is Professor of Sociology at the Institute of American Studies and Polish Diaspora of the Jagiellonian University in Kraków, Poland. Previously, she was Fulbright Scholar at the New York University. Her main research areas are migration and gender.

Branislav Radeljic is Senior Lecturer in International Politics at the University of East London. His forthcoming books examine European Community involvement in the Yugoslav state crisis and the role of non-state actors (London: IB Tauris) and Europe and the post-Yugoslav space (Farnham: Ashgate).

Francesco Ragazzi is Assistant Professor in International Relations at Leiden University (Netherlands) and Associate Researcher as the Center for International Studies and Research (CERI), Sciences Po Paris (France).

Sean P. Roberts is a Visiting Researcher at the Norwegian Institute of International Affairs (NUPI) in Oslo. His research interests focus on political development and transition across the post-Soviet space. His recent publications include the monograph *Putin's United Russia Party* (Routledge, 2012).

Oana Romocea is a doctoral candidate at the School of Cultural Studies and Humanities, Leeds Metropolitan University. Her research interests concentrate on the intersection of media studies, migration and transnationalism.

Rita Sanders is Lecturer in Social and Cultural Anthropology at the University of Zurich. She has conducted research in Kazakhstan and Germany. Her thesis and several articles deal with migration, ethnicity and the construction of home.

Ulrike Ziemer is Lecturer in Sociology at the University of Winchester. Previously, she has worked as Postdoctoral Research Fellow at the Centre for East European Language Based Area Studies (CEELBAS), School of Slavonic and East European Studies (SSEES), University College London. Her publications include the monograph *Ethnic Belonging, Gender and Cultural Practices: Youth Identities in Contemporary Russia* (Stuttgart: ibidem Verlag, 2011).

Acknowledgements

This edited volume is the outcome of a two-day international workshop on diasporas and cosmopolitanism in Central and Eastern Europe funded by the Centre for East European Language Based Area Studies (CEELBAS) and held at the School of Slavonic and East European Studies (SSEES), University College London. The workshop brought together an interdisciplinary group of scholars to discuss the notion of cosmopolitanism in the context of the post-socialist space, diasporas and migration. The workshop allowed these scholars to share their diverse research findings related to cosmopolitanism, diaspora and migration in the post-socialist context. We are very grateful to CEELBAS, in particular to Professor George Kolankiewicz, Dr Robin Aizlewood and James Perkins, for their support in the realization of this workshop. Furthermore, we would like to thank all the workshop participants for making this edited volume possible.

Introduction

East European diasporas, migration and cosmopolitanism

Ulrike Ziemer and Sean P. Roberts

The breakup of the Soviet Union in 1991 and the ensuing collapse of socialist regimes comprising the former Eastern bloc changed the geographical map of Central and Eastern Europe (CEE). Not only did whole nations disappear and (re-)emerge, but migration to, from and within this region significantly increased. For example, the political, economic and social transformations occurring in the late Soviet period, together with numerous ethnic conflicts, created sweeping migration flows estimated at around 25 million people (Korobkov 2008: 69). As with the former Soviet Union, states within the old Eastern bloc were not excluded from this upsurge in migration, with the ethnic conflict in the former Yugoslavia alone producing more than two million refugees and internally displaced persons. Although more than 20 years have passed since the collapse of the Soviet Union, these immense political and social transformations have left a trail of unanswered questions relating to citizenship, resettlement, revived diasporic transnational belonging in host and home societies, as well as ensuring the continuing importance of identity politics in this part of the world in general.

In this edited volume, we explore some of the many different facets of diasporic life and migration across CEE by specifically employing the concept of cosmopolitanism. At present, cosmopolitanism presents a number of problems for researchers, appearing in social science debates as a somewhat elusive concept with a number of different meanings (Beck and Grande 2007). In this volume, cosmopolitanism is explored as a means to explain identity choices, outlooks and political practices. Based on empirical studies we aim to examine the ways in which the concept of cosmopolitanism can help us account for migration and diaspora politics in CEE. In this way, our edited volume complements recent publications which attempt to understand cosmopolitanism as a product of a particular context, rather than the convergence on, or adoption of, a distant European model (Clifford 1998; Grant 2010: 125; Mandel 2008; Werbner 2008a).

Fine (2003: 432) identifies a 'new actually existing cosmopolitanism' which not only has flourished since 1989 but also has spread throughout the social science literature. In recent years, a large part of this renewed interest has focused on broadening the view of cosmopolitanism while drawing a new map of its distribution. Anthropologists have been at the forefront of this focus. Werbner (2008a: 1), for example, maintains that 'both in practice and substantive terms, a situated cosmopolitanism may indeed be at the heart of the discipline of anthropology'. Conversely, Ashis (1998: 146) stresses that 'Europe and North America have increasingly lost their cosmopolitanism that considers Western culture to be definitionally universal and therefore automatically cosmopolitan'. Yet, previous studies have shown that people living in non-Western cultures often have to negotiate and deal with more than one culture (Piot 1999). In line with these thoughts and insights, this volume adds to the 'map' of diverse cosmopolitanisms, so to speak, by extending the existing discussion to the post-socialist context. This discussion, with a few exceptions (Grant 2010; Humphrey 2004; Richardson 2006; Ziemer 2011), has been largely notable for its absence in existing literature.

The purpose of this Introduction is to familiarize the reader with the philosophical roots and scholarly development of the concept of cosmopolitanism and the different ways in which it has been applied in social science debates. The first part of this Introduction begins by tracing the concept of cosmopolitanism from its historical roots to its application in the present. The second part discusses the concept of cosmopolitanism as an alternative to the concept of diaspora. In addition, the second part draws attention to the post-socialist condition and its importance for analyses of present-day societies and politics in this part of the world and debates on cosmopolitanism in general. The Introduction concludes with an overview of the structure and chapters comprising this volume.

Tracing the concept

Cosmopolitanism has its roots in the Greek philosophical school of Stoicism. The Stoics believed in the ideal of the 'cosmopolis' or 'world state', to which all human beings belong, and whose tenets must be realized in action, regardless of local conditions (Toulmin 1990). The Stoics also believed that they were not removed from local affiliations, but that they were surrounded by a 'series of concentric circles' (Nussbaum 1996: 4), each considered to represent a different kind or level of attachment or identification. These circles began from 'the self' and extended outwards to include family, group, city and country and then to humanity as a whole (Vertovec and Cohen 2002: 12). The Greek Stoics' idea of a 'world citizenry' was revived by Kant in *Perpetual Peace* (1795), where he argued that world peace can only be achieved by a cosmopolitan state founded upon the otherness of the other. Kant also introduced the concept of 'cosmopolitan law', according to which individuals have rights, but as 'citizens of the earth' rather than as citizens of particular states (Kant cited in Reiss 1970: 97). In other words, the 'cosmopolitan' is a citizen of the world, an integral part of the world (Nussbaum 1996).

Accordingly, from its very first appearance in social science debates, the concept was associated with an orientation to a larger community of people, regardless of their location. For example, cosmopolitanism was introduced to sociology in order to distinguish those people 'oriented significantly to the world outside' (cosmopolitans) from those with dominant local orientations (locals)

(Merton 1964 [1949]: 393 cited in Pichler 2010). Thus, it could be argued that Merton from the outset acknowledged the complexities and contradictions of the concept of cosmopolitanism (Pichler 2009: 706), while hinting at its 'dialogical imagination' or, as Beck (2002a: 18) frames it, the clash of cultures and rationalities within one's own life – the 'internalized other'.

Beck himself employs the idea of a cosmopolitan outlook, which is exemplified by a reflexive engagement with both sameness and difference, opening up useful ways to investigate personal and cultural levels of cosmopolitanism. A cosmopolitan outlook is characterized 'by reflexive awareness of ambivalences in a milieu of blurring differentiations and cultural contradictions' and contains 'the possibility of shaping one's life and social relations under the conditions of cultural mixture' (Beck 2006: 4). At the same time, Beck portrays the cosmopolitan world as a 'glass world', one in which:

> Differences, contrasts and boundaries must be fixed and defined in an awareness of the sameness in principle of others. The boundaries separating us from others are no longer blocked and obscured by ontological difference but have become transparent.
>
> (Beck 2006: 8)

Accordingly, in the current age of migration, social reality cannot be simply described in terms of a sharp differentiation in 'similar' and 'different' (or 'us' and 'them'). Although migration is nothing new, it has increasingly acquired a new character, one which challenges the sovereignty of states. In addition, and with the appearance of 'transnationalism' and new mobilities, important, durable and simultaneous relationships of a political, economic, social or cultural nature in two or more societies have been established. In this way, people are no longer surrounded by a sole culture and cultural differences are no longer 'subsumed into a universalism, but are accepted for what they are' (Beck 2006: 57). In other words, today's everyday 'reality' is no longer constituted by a dichotomy of the universal versus the particular, because the universal itself appears as a particular form (Ziemer 2009: 412). Yet, despite the plurality of cultures in people's everyday lives, localism has experienced a revival. For this reason, cosmopolitanism serves as a prism through which one can account for the particular, for the intermingling of local, regional and national affiliations and identities in one's everyday life. Hence, the argument is that cosmopolitanism cannot exist without localism (Beck 2002a; Hollinger 2001).

Following this reasoning, sociologists and anthropologists invoke the notion of cosmopolitanism to describe diversity on a more immediate individual level. Hollinger (1995: 86) contends that 'cosmopolitanism is more oriented to the individual, whom it is likely to understand as a member of a number of different communities simultaneously'. Cultural approaches to cosmopolitanism were primarily pursued by Hannerz, who views cosmopolitanism as a perspective and state of mind as well as a mode of managing meaning (Hannerz 1990: 238). Hannerz also suggests that 'the perspective of the cosmopolitan must include relationships to a plurality of cultures', which he understands as 'distinctive entities' (Hannerz 1990: 239). In other words, cosmopolitan practices imply an acknowledgement of one's own culture alongside an active engagement with other cultures.

In political science debates, the concept of cosmopolitanism has been invoked to examine global processes of democratization in which transnational political institutions, such as the United Nations and the European Union, transcend the nation-state in governance (cf. Archibugi 2003). The European Union, for example, is described as 'the first international model which begins to resemble the cosmopolitan model' (Archibugi, 1998: 219). In this respect, European integration is an attempt to generate cooperation between different and often contradictory individual interests, groups and member states, where boundaries are constantly negotiated (cf. Calhoun 2002; Rumford 2005). Furthermore, cosmopolitanism has also been discussed in the context of global civil society, exercising democratic expression in an emerging transnational public sphere (Köhler 1998; Robinson 2008). These transnational bodies, however, do not by themselves constitute a state, but are able to address policies which cut across national borders, such as crime and pollution (Kaldor 1996). Critics, such as Calhoun (2002: 90), claim that these developments reflect a certain Western-centric discourse or perhaps a new kind of colonial practice.

Despite the increased interest explaining 'new' global realities, cosmopolitanism has suffered from being presented as an elite-driven phenomenon (Friedman 1995). Many studies have discussed a 'global' cosmopolitan elite 'who possess the knowledge and skills that currently fit productively with economic transformations engendered by rounds of globalization across cutting-edge, emerging industries' (Kanter 1995 cited in Skrbiš *et al.* 2004: 119). For Hannerz (1990: 245) these mobile elites remain 'metropolitan locals' instead of becoming fully cosmopolitan. They are either citizens of First World countries or they belong to the privileged classes of non-Western countries and identify with Western ideals (Skrbiš *et al.* 2004: 120). At the same time, there are those individuals who stand in stark contrast – those who are confined to place and are locally rooted (Urry 2000) or those who are the victims of modernity, deprived of customs of national belonging (Pollock *et al.* 2000). For Pollock *et al.* (2000: 582), refugees, diasporic peoples, migrants and exiles are those who comprise the cosmopolitical community.

A large part of the debate on cosmopolitanism focuses on questions of citizenship. In the past, broadly speaking, there have been two competing social science approaches used to discuss issues of state citizenship and membership. First, citizens appear as active participants in the public affairs of the polis (Turner 1990). Such a view is based on classical civic traditions stemming from an Aristotelian picture of 'man as a creature born to fit society' and from Rousseau's notion of the social contract (Dickinson and Andrucki 2008: 101). Secondly, with the rise of market-oriented society, the classical 'active' obligation-based civic ideal was gradually replaced by a modern 'passive' or 'liberal' ideal with origins in 'bourgeois' values of the cities of early modern Europe (Burchell 1995; Dickinson and Andrucki 2008: 101). However, new forms of citizenship have become possible in an interconnected and globalized world. People, including migrants, diasporic people, as well as locals, find themselves 'belonging' to more than one community and also to more de-territorialized and transnational communities. Citizenship in a global age implies the possibility of belonging to the world as a whole, rather than to nations or geographically bounded communities (Molz 2005: 518). Yet, even though cosmopolitan citizenship transcends the boundaries of the nation-state, it does not imply that the meaning of national belonging disappears. Instead, as many theorists have demonstrated and as subsequent chapters show, cosmopolitan and national identities are intertwined and overlapping (Appiah 1998; Cheah 1998).

Cosmopolitanism and diaspora

Whereas the previous discussion traced the scholarly development of the concept of diaspora, this section sheds light on cosmopolitanism as an alternative to the concept of diaspora. As indicated in the previous section, in the twenty-first century diasporic identities have become mobile, while fixed origins have become questionable. Yet, issues of homeland and diaspora are still important, although this relationship is increasingly more complex than a simple dual attachment. Hence, diasporic experiences can be viewed from a cosmopolitan perspective that offers enough flexibility to describe the diverse experiences and political practices.

Many of the issues raised in debates on globalization are echoed in Beck's work on cosmopolitanism and its implications for sociology. Beck stresses a process of cosmopolitanization that counteracts globalization, identifying it as a methodological concept which has the power to provide an alternative image of social life, one which seeks to comprehend the otherness of nature, other civilizations and other modernities. Cosmopolitan sensibilities are not merely something taking place 'out there', generated through greater global mobility and interconnectedness; rather, cosmopolitanism has to be viewed 'from within' (Beck 2002a). Thus, in view of shrinking time and space distances under global processes, individuals from diasporic communities are often understood as having dual attachments or, more recently, a multitude of cultural attachments (Ali 2003; Bhabha 1990; Bhimji 2008; Marsden 2008). Gilroy (1987), for example, alluding to Du Bois's notion of 'double consciousness', describes a kind of duality of consciousness amongst individuals from diasporas. For him, diasporic individuals are decentred in their attachments – they are simultaneously 'home away from home' or 'here and there', or British and something else. Similarly, Clifford (1994: 322) proposes that 'the empowering paradox of diaspora is that dwelling here assumes a solidarity and connection there … [it is] the connection (elsewhere) that makes a difference (here). But there is not necessarily a single place or an exclusivist nation'. In other words, social reality in a globalized world does not necessarily imply an absence of belonging but rather the possibility of belonging to more than one ethnic and cultural locality simultaneously (Bhimji 2008; Werbner 1999; Ziemer 2011).

Accordingly, the traditional concept of diaspora has been criticized for conveying essentialism that does not exist under global conditions. Diasporic forms flourish and ethnicity is replaced by hybridity (Clifford 1997) or a 'third space' (Bhabha 1990) only known to diasporic people. Critics stress cross-fertilization between different cultures as they interact, leading to creolization (Cohen 2007) or 'new (hybrid) ethnicities' (Hall 1996). In recent studies on cosmopolitanism, it is shown that diasporic people familiarize themselves with other cultures and know how to move easily between them (Werbner 1999: 20). Nonetheless, in the debates on cosmopolitanism and diaspora it is emphasized that this might not always be the case. Cultural diversity can also lead to the mythologizing of traditions (Shukla 1997) or what Beck (2006) terms 'as if' nationalisms that oppose any cosmopolitanization of one's diasporic culture.

The concept of diaspora has been used for displaced people, migrants and transnational peoples. Yet, in its current definition it seems to be out of touch with contemporary diasporic social life. Even though notions of 'homeland' are a major pillar for diasporic being, as Darieva's Chapter 2 shows, the meanings of 'home/homeland' have changed for diasporic people (cf. Ali 2003; Falzon 2003; Ziemer 2009, 2011). Closely linked to the notion of diaspora and home is the idea that diasporas maintain a collective historical memory, constituting a subjective definition of ethnic belonging – a theme taken up in Chapter 6 in this volume, which explores changes in the collective memory of migrant Poles living in Berlin. Appadurai and Breckenridge (1989: i) argue that 'diasporas always leave a trail of collective memory about another place and time and create new maps of desire and attachment'. In general, a diaspora can, to some degree, be held or recreated in the mind, through cultural artefacts and through a shared imagination (Ziemer 2011: 44). In this way, Cohen (1996: 516) stresses that 'identification with a diaspora serves to bridge the gap between the local and the global'. The complexity of cultural identification and the consequent awareness of multi-locality amongst diasporic peoples stimulate a constant process of formulating and reformulating diasporic representations. Hence, for Hall (1990), diaspora is composed of ever-changing representations which provide an 'imaginary coherence' for a net of flexible identities. It is these diasporic representations that are no longer fixed in an apparent age of globalization.

Locating cosmopolitanism in the post-socialist context

As the chapters in this volume show, the post-socialist context is vital for exploring the diverse aspects of cosmopolitanism. Studies emerging in the first decade after the collapse of communism employed post-socialism as a purely descriptive category, restricting analysis to the state level of political and economic change (cf. Malia 1994; Stiglitz 1994). This approach portrays 'transition' in post-socialist regions as an example of the way in which 'the micro' is determined by or an expression of macro structures (Burawoy and

Verdery 1999). However, as indicated in a number of sociological and anthropological studies, everyday experiences are rather characterized by innovative responses that reveal the influence of the socialist legacy in some way or another (cf. Kay and Kostenko 2008; Mandel 2002; Mamattah 2008). Although much of this literature stresses the specificities and peculiarities of the post-socialist context, these studies also show that everyday experiences cannot be viewed as completely different from the West. Instead, many of the issues discussed transcend national boundaries and artificial 'East-West' divisions (Kay and Kostenko 2008). This, however, does not mean that people's everyday experiences in post-socialist regions and their responses to societal and political changes can be generalized. Instead, they have to be situated within the particularities of place, specific histories, cultural and socio-economic conditions. By considering the socialist past in an analysis of the post-socialist present, we can better highlight the tensions and ambiguity that influence people's present experiences. Therefore, the concept of cosmopolitanism may have its limitations when it comes to researching and understanding minority groups in post-socialist countries. Yet, this does not mean that cosmopolitanism has to be abandoned completely. Instead, in such cases cosmopolitanism can highlight the different ways in which groups of people form their diverse sense(s) of belonging by a selective and diversified engagement with the socialist past.

Before the collapse of communism, one could argue that cosmopolitanism or internationalism was something concrete in Eastern bloc countries – something that could be felt (Grant 2010). There were universalist tendencies, even though they were selectively applied. In view of the premises of socialist and communist ideologies, one could assume that cosmopolitanism is a natural outcome of these ideologies. However, history showed a somewhat different development. While cosmopolitanism and cosmopolitan thinking were encouraged in the West and not openly opposed, political practices in communist Europe often attempted to eradicate any free cosmopolitan thinking or cosmopolitan practices that did not conform to socialist and communist ideology. Generally speaking, the socialist model promoted 'national' and 'cultural' differences, but at the same time political unity. The utopian Marxist-Leninist ideology that envisioned the abolishment of any form of affiliation beyond a universally shared class-consciousness appeared like a supra-ethnic ideology that sought to supersede particularistic tendencies (Humphrey 2004). It reflected a simple use of the term 'solidarity', in particular with the working-class struggles in other countries (Grant 2010: 132). Marxist-Leninism urged proletarians of the world to unite, and to shed the fetters of false consciousness that bound them to a place, race, language and religion (ibid.).

However, it became obvious in the Soviet period, in particular under Stalin, that not all universalist tendencies were desirable. So-called 'rootless persons', or persons with doubtful or multiple loyalties, were often accused of being traitors to the motherland. Jews, for example, were a main target group, although other migrant groups such as Armenians and Germans were also targeted. As a group, Jews were perceived as less than national and thus insufficiently attached to the nation. On the other hand, they were more than national and hence threatened the nation's transcendent, universal status (Kofman 2005: 89). According to Soviet ideology, minorities should have been assimilated and indeed by the end of the Stalinist period many had adopted Russian language and culture. In short, whereas the utopian Marxist-Leninist ideology espoused a policy of internationalism that, to some extent, led to cosmopolitan practices (a subject analysed in detail in Chapter 5 in this volume), the term *kozmopolitizm* became a derogatory category, implying a notion of superiority and categorization that was counter to official Soviet thought (Humphrey 2004; Kofman 2005).

The structure of the book

This volume explores the relationship between East European diasporas, migration and cosmopolitanism situated in a specific localized context in the post-socialist space. In this way, the structure of this book is determined by specific diasporas and/or locations instead of broader themes involved in the study of diasporas, migration and cosmopolitanism. In Chapter 1, Rebecka Lettevall explores the case of the Nansen passport, perhaps an early example of the practice of cosmopolitanism. In the immediate aftermath of the First World War, with the collapse of Russian, Ottoman and Austro-Hungarian empires, Europe experienced large numbers of 'stateless' people and cross-border migration. In 1921, and in order to alleviate this situation, the so-called Nansen passport was introduced. By 1942, they were in use in 52 countries, with over 450,000 issued. Lettevall explores the case of the Nansen passport in order to determine whether it could be seen as an early example of cosmopolitanism in practice.

In Chapter 2, Tsypylma Darieva continues the historical theme central to Part I of this book by focusing on the American-Armenian engagement with their so-called 'historical homeland'. Armenians were included in this edited volume owing to the sheer scale of their international presence. At the time of writing, the number of Armenians living outside of the country is estimated to be at least 10 million, making the Armenian diaspora one of the largest, as well as one of the oldest. Darieva examines the long-standing relationship between diaspora and homeland and the ever-changing meanings of homecoming – all of which are considered key features of diasporic identity. Her chapter is based on ethnographic observations in the Boston area (MA, USA) and in Armenia, where she conducted fieldwork to explore diasporic rhetoric and transnational activities, mostly between the USA and the Republic of Armenia.

In Part II of this volume, attention shifts to diasporas and diasporic politics in the former Yugoslavia. Although the disintegration of the Socialist Federal Republic of Yugoslavia was acknowledged by the European Community (EC) on 15 January 1992, drafting the policy of recognition was problematic, mainly because of intermittent concerns that granting independence to the republics of Slovenia and Croatia might trigger further violence in the remainder of the Yugoslav Federation. Within this context, Branislav Radeljic (Chapter 3) analyses political activism among members of the Slovenian diaspora, in particular their promotion of Slovenian secessionist interests and subsequent influence over EC policy. Cosmopolitanism implies an understanding of difference, as well as universal inclusion – two aspects which are often promoted by elites and that acquire an increasingly important role in times of crisis. In the case of Yugoslavia, the EC policy of recognition can certainly be interpreted as European support for independence from certain specific (Serbian) authorities. In presenting a detailed analysis of diaspora activism during the Yugoslav crisis, the author makes extensive use of original sources that have not been used in existing literature on the subject of the European Community and Yugoslavia.

In Chapter 4, Francesco Ragazzi continues the focus on diaspora politics in the former Yugoslavia. Much of the transnationalism literature in the 1990s celebrated the discovery of a new sociological object, what was termed 'transnational communities'. These communities, which were often conflated with 'diasporas', were described as the new social form of the twentieth century, challenging the nation-state, holding a promise of cosmopolitanism and post-national belonging. In his chapter, Ragazzi argues that much of this enthusiasm has been exaggerated, and is in fact contradicted by sociological and anthropological empirical studies. His central argument is that while diasporic discourses and practices of citizenship question and redefine the nationalist relationship between identity and territory, they none the less operate as a tool to separate an 'inside and outside' and an 'included and excluded'. As a result, post-territorial practices can be read as the constitutive 'other' of practices of post-nationalism and cosmopolitanism. Through the empirical study of the post-Yugoslav states and Croatia in particular, Ragazzi explores the way in which diasporic citizenship operates as a technology of securitization of a particular (official) ethnicity, identified as 'transnational' at the expense of the unwanted ethnicities still residing in the territory, functioning therefore as a mechanism of post-territorial nationalism.

Part III examines border-crossing from Eastern to Central Europe and vice versa, focusing on questions of diaspora and cosmopolitanism in and beyond Germany. In Chapter 5, Rita Sanders explores cosmopolitanism in the context of the German diaspora in Kazakhstan. This chapter investigates the interplay of cosmopolitanism and mobility through the example of Kazakhstani Germans, who either moved to their 'historic homeland', Germany, or stayed in Kazakhstan. Sanders explores Kazakhstani Germans' everyday practices and attitudes toward the other by elaborating on rootedness and openness in terms of primordial diaspora thinking and Soviet-style internationalism. In this context, cosmopolitanism is understood as a capacity to build relationships and identities of openness, despite cultural differences.

In Chapter 6, Dorota Praszałowicz focuses on Germany itself, providing an analysis of Polish migrant life in Berlin by exploring migrants' memoirs and confronting memories of the 'old' and 'new' Polish migrants in Berlin. The term 'old migrations' is used to denote Polish migration to Berlin from around the middle of the nineteenth century until the late 1930s, while 'new migrations' refers to those Poles who found themselves in Germany at the end of the Second World War, as a result of forced labour migration or displacement. New migrations also include Poles who migrated to Berlin in the 1970s, 1980s and 1990s. The author shows that old migration memories are embedded in the Polish national discourse, focusing on Polish ethnic community-building and cultural continuity, while resisting assimilation processes. Contrary to the 'old' migration discourse, the 'new' migrants' narratives are a product of a globalizing world in which multiple national identities and transnationalism are perceived as natural.

In Part IV of this volume, ethical considerations surrounding researching diasporas and migration are considered. Ethical issues impact on all forms of social research. Researchers have a set of moral principles that concern key issues such as confidentiality, anonymity, legality, professionalism and privacy in relation to the research process. During and after fieldwork, researchers may encounter ethical challenges that were not anticipated. These issues are often even more pressing when research is conducted with migrants and by migrant scholars. Darya Malyutina discusses the ethical difficulties she encountered during ethnographic fieldwork among Russian-speaking migrants in London. Her chapter seeks to discuss two major aspects of ethnographic fieldwork: access to the field and relationships with research participants. The author explores a number of pertinent issues, including dealing with distrust and rejection as well as sustaining relationships with respondents during fieldwork. This last point is particularly relevant, as the ethnographic research on which this chapter is based examines friendships and informal relationships amongst Russian-speaking migrants in London.

In Chapter 8, the final chapter, Oana Romocea discusses ethical issues arising from research amongst Romanian migrants settled permanently or temporarily in the UK. Following the fall of communism in December 1989, the Romanian diaspora in the UK saw its numbers steadily rise, although Romania's EU entry in January 2007 resulted in a much more significant increase in UK-bound migration. Recent migration scholarship has called for a methodological shift to facilitate the study of such sizeable trans-border flows. Until recently, the nation-state provided the predominant analytical framework for understanding contemporary migration movements across political borders, but this macro-level approach has its shortfalls. Among them is the problem of effectively engaging with the micro-level, everyday-life experiences of migrants and in understanding how daily experiences affect migrant identity and outlooks in the host society.

Overall, we hope to present cosmopolitanism as a useful conceptual framework to study migration and diasporas in the post-socialist context. The material presented in these chapters points to a continuing relevance for post-socialism as an analytical category, although this category does not detract from the suitability of cosmopolitanism as a means to explain political and societal developments concerning migration and diasporas in this part of the world. It is our hope that the material offered in this volume will generate rich discussion among area studies scholars, but also among researchers working in other social science disciplines as well.

Part I

The past in the present

Fostering cosmopolitanism

1 Cosmopolitanism in practice

Perspectives on the Nansen passports

Rebecka Lettevall

The end of the cold war brought many changes within not only the political but also the social, cultural and conceptual understandings of Europe. In response, academic scholarship attempts to explain these changes and challenges Europe faces by re-invigorating the centuries-old concept of cosmopolitanism. Yet, this does not mean that the age-old idea of a world citizenship with roots in ancient Greece was simply transferred to the new post-cold war context. Instead, the concept of cosmopolitanism has been filled with new content, and thus what is usually referred to as *new cosmopolitanism* has been born. The revised understanding of the old concept and a revised view of a world citizenship and globalization point toward the question to what extent it is possible to practice a cosmopolitan citizenship.

After the end of the First World War, Europe faced many changes, just as after the end of the cold war. These changes saw the appearance of large numbers of stateless people, not least in Europe. In order to solve this situation, the idea of the Nansen passport was born, which in the inter-war period was a widely known and discussed document, but is more or less forgotten today. In this chapter, I discuss the concept of cosmopolitanism and how it is used, or not used. I present an analysis of the Nansen passport and explore whether it might be called cosmopolitanism in practice. Studying the history of concepts like cosmopolitanism is more than just the study of something in the past. It confronts us with the historicity of our own concepts, and makes us aware that our own premises are changeable, historical matter.

Migration in the inter-war period

Migration is an integral part of human life and experience. However, after the end of the First World War, the migration situation in Europe imposed challenges that transgressed the level of nation-state. The number of refugees and the causes of migration were of a scale that affected not only the new but also the old residential countries of refugees. One reason for this was the dissolution of the former Russian, Austro-Hungarian, German and Ottoman Empires, which resulted in the re-drawing of the map of Europe, including the creation of new states in Eastern and Central Europe such as Estonia, Latvia, Lithuania, Poland, Austria, Hungary,

Czechoslovakia and Yugoslavia. This great transformation from multicultural empires to nation-states displaced many members of ethnic minorities from the newly formed states. This forced millions of people to leave their home countries. Some of them joined their ethnic group in other states, while others became refugees. Noteworthy here is that the term 'refugee' was first used for the French Huguenots who left France after the Edict of Nantes in 1695. The development of the nation-state was connected to a change of view among minorities: suddenly they no longer fitted the state (Zolberg 1983). Some became stateless, no matter whether they had left their country or not.

The global situation was changing too, and from the beginning of the twentieth century, several non-European countries were less willing to host new immigrants without any demands on them. At the same time, as many people lost their citizenship, they also lost their right to work or find dwelling, at least in legal ways, as they did not have proper documents of identification. According to Hannah Arendt, this implies a 'fundamental deprivation of human rights' which makes us become aware of the existence of 'a right to have rights' (Arendt 1976: 296). The disillusioned atmosphere after the First World War and a widely witnessed sense of loss – not only of human lives and material things but also of ideals, hopes and culture – became evident in the masses of stateless persons and refugees. As the war had changed the map of Europe considerably, millions of men, women and children had lost their identity as citizens and as a consequence had even lost their rights as citizens of one country. The food supply was very poor and caused horrifying starvation, especially in Russia and Ukraine. Newspapers of the inter-war period testify to the incredible amounts of refugees from the 'old' Europe of former empires as the 'new' Europe was organized in nation-states. This indicates that the refugee question was also an issue in the public sphere. Homeless refugees seem to have been visible in more or less every city in Europe. The official number of refugees at this time varies significantly, but there seems to have been several hundreds of thousands, or even millions.

The development of nationalist doctrines in the new Europe contributed to the refugee situation (Skran 1995: 24f). The development of several national social welfare programmes, ironically enough, made some countries less willing to accept new habitants. Newcomers were also considered as a potential political threat, since they had already shown patriotic disloyalty by having left their country and so could be seen as untrustworthy. Skran (1995: 29) points out that the mass refugee movements of the inter-war period were by-products of the political efforts to create ethnically pure nation-states and ideologically homogeneous political systems.

These refugees needed a legal document. Before the First World War, there were only a few refugees who had a vague legal status, but after the war, thousands of refugees crossed borders in Europe without any legal protection. What was previously acceptable now proved to be inadequate for properly responding to the huge new demands. The legal anomaly of these refugees was identified as a problem by many legal scholars, who tried to solve it in different ways (Grahl-Madsen 1983). A person without citizenship or a legal document was unprotected and had no rights. In this context, it is worth mentioning that Russian refugees faced the worst situation, as they had no new state that would accept them as citizens after the October Revolution in 1917. In this situation, with many states disappearing and new ones arising, the idea of a solution with a new kind of system of identification was born: particularly responsible for its birth was Fridtjof Nansen.

Fridtjof Nansen

The Norwegian scientist, explorer and diplomat Fridtjof Nansen (1861–1930) was the most important entrepreneur behind the idea of Nansen passports and so they were named after him. He is a person who could be easily depicted as a true hero of our times for several reasons. He was tall, blond and handsome and reached fame after his adventurous expedition with the ship *Fram* (1893–95), which held him caught in the icy north of Russia instead of taking him to the North Pole (Huntford 1997). However, this served as a proof for his theory of the Arctic Ocean streams. He was the first European to cross Greenland from east to west – an experience that he often returned to later. After his scientific expedition and career, he entered the political and diplomatic world. First, he took an active part in the debate about the dissolution of the Swedish–Norwegian union, arguing for the freedom of Norway (Huntford 1997: 393–414). After the separation from Sweden in 1905, Nansen served as Norway's ambassador to Great Britain (1906–08). During the war, he negotiated with the USA on behalf of Norway in order to secure a grain supply. After the war, he was the chair of the Norwegian Association for the League of Nations, a position he held for the rest of his life. Through his diplomatic experience he became acquainted with many of the people who would turn out to be important for the future. His efforts were awarded with the Nobel Peace Prize in 1922 and the Nansen International Office for Refugees was awarded the same prize in 1938 (www.nobelprize.org).

Had it not been for some problems within the League in finding a unifying and acceptable candidate for the post of High Commissioner for Refugees, Nansen might never have been appointed (Marrus 1985: 87–90). But as his diplomatic career had brought him into contact with humanitarian projects where the problems of Russia were in focus, it proved he had very good experience for the tasks of High Commissioner. Two huge projects are particularly worth mentioning here, as to some extent they demanded a solution to the refugee question. In the 1920s, he was involved in a project to repatriate prisoners of war from Russia and a project to prevent the famine and its consequences in Russia and Ukraine through international aid programmes (Vogt 2007). In short, his suitability for the post of High Commissioner was his persona, including his scientific and explorer background, the ethical credibility he had achieved through his work on the two Russian projects, and that he was a citizen from a neutral state (Lettevall forthcoming).

The Nansen passport

Passports, as documents issued by a national government to identify the holder in terms of name, date and place of birth, sex and, usually, as a national citizen, and in some cases even with a self-defined ethnicity, had generally, and perhaps unexpectedly, lost their importance with the rise of nation-states and nationalism in the nineteenth century. Since the French Revolution, passports or similar documents of identification have been an essential and legitimate means for the state's monopolization of control over the movement of people. During the relative peace in Europe in the nineteenth century, the amount of travelling increased, not least because of the technological development of steam trains and steam ships, which almost caused the European passport system to break down. However, during the First World War, this passport system was re-established, together with other restrictions of movement (Torpey 2000: 57–143).

In this context, the Nansen passport was created fairly quickly. Fridtjof Nansen was appointed by the League of Nations in 1922. In March, he drew the League's attention to the difficulties of Russian refugees in migrating and suggested that they should be provided with some kind of travel and identity document. In July 1922, Nansen had an intergovernmental conference on the issue, a conference that ended up with an Arrangement (of 5 July 1922), that is an agreement within the League of Nations that is not legally binding, but rather to be considered as a recommendation for the signing states. The Arrangement introduced the Nansen passport, an identity certificate for Russian refugees. Although this Arrangement introducing the Nansen passport was not legally binding, many states quickly accepted this Arrangement. Initially, the Arrangement was accepted by 16 countries, but soon it was accepted by many more governments, with a peak in 1929 of over 50 states introducing the Nansen passport. Noteworthy here is that this Arrangement was soon extended to include the large group of Armenian refugees in 1924 and thereafter even more refugees were included (Marrus 1985: 74–81).[1]

Holders of a Nansen passport could travel and were also issued identity cards, which increased their chances of employment and decent dwelling. Nonetheless, the Nansen passport did not guarantee opportunities for work or home, but at least it made them legally possible. Not surprisingly, this very first international refugee agreement had its shortcomings. First, it was primarily directed at Russian refugees, even though there were many ethnically other refugee groups. Second, the freedom of travel was in some ways limited, as passport holders were not always allowed to return to the state of issue, once they had left the state. Finally, on many occasions, there was a fee which had to be paid annually for the renewal of the passport and many passport holders could not afford this fee.

Among the European states that had accepted the Agreement and issued Nansen passports, there were high hopes that the League of Nations would be able to solve the refugee question. Nevertheless, it soon became obvious that despite this Agreement, the issuing of Nansen passports and identity documents was not organized in the right way. The members of the League of Nations did not want to recognize the Soviet Union, and the Soviet Union refused to cooperate with the League of Nations. In addition, the Nansen International Office for Refugees had neither enough funding nor full authority. The League of Nations financed some of the administrative costs of the Nansen International Office for Refugees, but it was mainly supported by private contributions and the yearly fees which were required for the renewal of the Nansen passports. The worldwide Depression of the 1920s also had an impact, as it resulted in high unemployment and growing difficulties for refugees and immigrants to find work and earn an income. As a result, the League of Nations had lost some of its credibility by the time new groups of refugees emerged as a result of the Spanish Civil War in 1936–39. Thus, these circumstances created a specifically difficult situation for the Nansen passport, although it survived a little longer until it completely vanished in connection with the Second World War.

The above discussion has established the background and the development of the Nansen passport and has shown that, despite its various shortcomings, the Nansen passport was a truly international refugee agreement adopted by many states, used to solve the refugee question. In the subsequent discussion, I explore the question of whether the Nansen passport could be seen as not only the first international refugee agreement but also as a kind of cosmopolitanism in practice.

About cosmopolitanism

Concepts have a history, and thus their meaning changes depending on the context. Some seem to be more changing than others. A large number of central concepts in modern politics, such as democracy, war and peace, risk being misunderstood by an ahistorical approach (Lettevall 2011: 179–88). Anyone discussing democracy today who turns to history for support can easily misunderstand Plato or Immanuel Kant, for example. Kant is often portrayed as an advocate of democracy, yet the form of government which he advocated was, more precisely, republicanism (Gerhardt 1995). To be sure, Kant's republic has some of the features we today associate with democracy, such as division of power and citizenship. At the same time, the concept of democracy in Kant's writings has a clearly negative connotation; democracy, according to Kant, is characterized by incompetence and disorder, and, as for Plato, threatens to descend into anarchy.

Another concept with a similarly long history is cosmopolitanism, which is explored in the context of the Nansen passports. Could Nansen passports be seen as cosmopolitanism in practice? In order to find answers to this question, cosmopolitanism is discussed through a reflective history perspective. For both Hans-Georg Gadamer and Reinhart Koselleck, language and concepts are central to how we understand the past. Gadamer's notion of *Wirkungsgeschichte* and Koselleck's project of conceptual history offer valuable insights into the historical study of ideas, insights which can move us beyond the simplified opposition between historical and ahistorical approaches to intellectual political history.

Any reflective conceptual-historical understanding which aspires to contemporary relevance will gain much from allowing itself to be guided by the idea of *Wirkungsgeschichte*. This perspective focuses on historical phenomena and the effects of tradition as well as their historical repercussions, and ultimately also the history of scholarship and academic research (Gadamer 1960). Koselleck's conceptual-historical approach opens up new possibilities for understanding both the past and the present. The changes in meaning exposed by conceptual-historical studies can reveal not only the past but also the present. The idea is that through discussing the Nansen passports, we might not only illuminate an almost forgotten initial practice international refugee politics but also receive a better understanding of cosmopolitanism.

In brief, cosmopolitanism has had three significant periods of importance. According to classic sources, the word originates from the saying of the Cynic Diogenes from Sinope; when asked where he came from, he answered, 'I am a cosmopolitan,' referring to the fact that he felt at home everywhere (Diogenes Laertius 1972 [*c*200 BC]). Nonetheless, it must not be forgotten that this quotation is problematic, as the source of it is Diogenes Laertius, who lived several hundreds of years after Diogenes from Sinope. The Stoics used the concept and integrated it in Stoic thinking, even if the Stoic understanding of it emphasized the moral universalist side of cosmopolitanism.

Some scholars have considered cosmopolitanism as a key concept of the Enlightenment (Brunner 1982; Kleingeld 1999). It was developed and influenced from the ideas of universalism and order. In particular, the philosopher Immanuel Kant (1724–1804) is one of the key figures for this development as he systematized cosmopolitan right and combined it with international right as well as with the idea of the state. According to Kant, people of the Earth belonged to one and the same realm, such that a violation of the rights of an individual was also a violation of humanity as a whole, no matter where on Earth, or against whom, the violation took place (Kant 1992 [1795]: 360). In intellectual history, this is an interesting milestone, as it illuminates the connection between the idea of human dignity with ideas of universal humanity and universal rights. However, Kant's idea builds on the idea of the nation-state rather than on the idea of the individual.

It is sometimes assumed that the interest in cosmopolitanism vanished at about the time of Kant's death. The general view is that the nationalism of the nineteenth century dominated the discourse in such a way that it was almost impossible to advocate cosmopolitan ideals. But that does not mean that the content of the concept was forgotten. Throughout the nineteenth century, and parallel with nationalist discourse, there were socialist and peace movements that stressed the *inter* aspect of the word 'internationalism'. Marx's writings are – in a multiplicity of ways – important in this respect, showing the doubleness of conceptual history. He condemned the concept of cosmopolitanism (a condemnation made not only by him) and the meaning of cosmopolitanism morphed into something akin to unwanted Jews. At the same time, he encouraged the internationalism that was so important in the nineteenth century within several popular movements. This is not to deny that the nationalist discourse was the dominating one. Even after the First World War, the impact of nationalist voices was substantial, although nations made more efforts to cooperate, such as with the founding of the League of Nations. This story is known at least on the surface by everyone who is interested in cosmopolitan theory.

Accordingly, the perspective of *Wirkungsgeschichte* is useful in order to clarify some different meanings of cosmopolitanism. When an old concept is used to explain a new setting it can lead to conceptual difficulties. No matter how the concept is defined or re-defined, it still carries meanings of its former uses that have become an inseparable part of it. This does not mean, however, that concepts have a life of their own. Concepts can only exist when they are employed and re-employed in new settings. One might say that a great deal of contemporary cosmopolitan theory in a sense breaks with the 200-year-long tradition of thinking about political theory from the perspective of the nation-state. Although analytical concepts are essential tools for any historian who wishes to give structure and coherence to the past, they can also obscure the particular historical context in question.

Cosmopolitanism in the inter-war period: an example

The status of the term 'cosmopolitanism' was not at its peak in the inter-war period. In fact, some saw it as a dangerous tendency, and even as an insult. An example of that is to be found in *the Encyclopedia of Sociology* of 1931, where the German sociologist Max Hildebert Boehm, in a Hegelian conceptual framework, describes cosmopolitanism as an 'abstract universalism' (Boehm 1931: 457–61). According to Boehm, it is necessary to take an 'identity bath' in order to switch from a natural individual to a cosmopolitan. It was a major transformation. But he also introduces what he calls a 'concrete universalism,' which would be the same as internationalism. This idea has its origin in the Romantic philosophy and is characterized by the levels between the individual and the world citizenship. Boehm stresses that these levels could be the family, the state, race, or the League of Nations, and emphasizes the importance of the state. Unlike cosmopolitanism, internationalism is not indifferent to the question of nationality. Even if the opinion of Boehm cannot be considered important, his view on cosmopolitanism coincides in general with what seemed to be the dominant opinion at that time. The words 'cosmopolitan' and 'cosmopolitanism' were not generally considered as too positive. Instead, after the war, the claim for internationalism in most cases was more accepted, which was shown in the attempts to organize a League of Nations. This indicates that even if cosmopolitanism was a word that generally would not be used in a positive sense in the inter-war period, the content of it could still exist, if one understands the word 'internationalism' as something that has to do with people transcending nations.

Cosmopolitan understanding

The strong interest in pursuing various forms of cosmopolitanism among scholars today should be seen, at least in part, as historically determined. The historical situation since the fall of the Iron Curtain has re-actualized this idea and created a lively debate in the wake of globalization. At the same time, the twentieth century was to a large extent characterized by a profound distaste for cosmopolitanism as a term and as a concept to be explored. Therefore, this period could be considered as a blind spot on the conceptual-historical map. In contrast, with the end of the cold war and when globalization took hold in Eastern Europe, it also contributed to new perspectives on cosmopolitanism. Proponents of new cosmopolitanism were quick to seek an affiliation with the historical tradition of cosmopolitanism by invoking Stoicism and Enlightenment cosmopolitanism, with Kant as a particularly important influence. Such efforts, however, sometimes lacked an understanding of conceptual history. Historical references were used more as decoration than as a conceptual resource for a more profound understanding of the present.

The development of contemporary cosmopolitan theories is often divided into three groups: roughly speaking, they are cultural, political and moral cosmopolitanisms. In addition, there is also an understanding of cosmopolitanism as a certain kind of individualism. This kind of cosmopolitanism might not be a system of ideas, but rather a person who refers to himself or herself, or is referred to by others, as a cosmopolitan, indicating that the person is an addicted traveller who feels at home everywhere. Such a cosmopolitan person could be described as a colonial person who travels around the world without really meeting anyone or interacting with the local cultures. This type has been called the tourist by Zygmunt Baumann and a globetrotter by Ulrich Beck (Lettevall 2008: 13–30).

Cultural cosmopolitanism claims the right of multiculturalism and opposes nationalistic ideals. This kind of cosmopolitanism encourages cultural diversity and, in this respect, has a connection to diasporas. It has grown through the increased migration and globalization after the end of the cold war. However, the Nansen passports are not cosmopolitan in this sense. The Nansen passports are not clearly cosmopolitan neither in the sense of a political cosmopolitanism. This kind of cosmopolitanism often has a Kantian background, advocating a world state or a world federation in which the nation-state alone is no longer a major political force. The globalized economy moves from one state to another, depending on where it finds the most fertile conditions and environmental pollution does not know national borders.

Political cosmopolitanism is sometimes applied to account for legal cooperation, such as the International Criminal Court (ICC). This kind of political cosmopolitanism could be used to shed light on the Nansen passports. Nansen passports were premised on specific features of political cosmopolitanism, as they were founded on collaboration between states and agreed on as an Arrangement by several states. In the inter-war period, the belief in the nation was very strong, and the League of Nations in a way both strengthened the nations and encouraged collaboration. The Nansen passports were not issued by the League of Nations, but by the nations themselves. This middle-level procedure caused some difficulties. For example, the passport holders were not granted civil rights and, thus, were second-class citizens.

Even moral cosmopolitanism has a background in Kant's philosophy. In *Perpetual Peace* (1795), Kant discusses political systems, but he also discusses cosmopolitan right, which appears to be a moral right of hospitality, founded on natural right. A person whose life is in danger has the right to visit any place on earth in order to save his or her life. But Kant also states that all persons are part of the same human realm, and that a violation of one person's rights is a violation against all of humanity:

> The growing prevalence of the (narrower or wider) community among the peoples of the earth has now reached a point at which the violation of right at any *one* place on the earth is felt in *all* places. For this reason the idea of cosmopolitan right is no fantastic or exaggerated conception of right. Rather it is a necessary supplement to the unwritten code of constitutional and international right, for public human right in general, and hence for perpetual peace.
>
> (Kant 2005: 84–5)

This quotation helps us to understand the Nansen passports. The inhuman living conditions of refugees demands action that goes beyond the legal systems and calls for the unwritten code of united humanity, with connections to the natural law tradition. The moral cosmopolitanism rests upon the idea of moral universalism: that all humans are equal, and not just your own family or compatriots. This has also been the issue for criticism, as there is a danger in judging others from your own perspective as well as implying that they would have the same values as you do (Appiah 2007).[2] If considered in this context, cosmopolitanism appears to have a weak meaning, at least in a legal sense, claiming only the dignity of universal mankind. Being human implies having a decent life, and also helping people who are not in your position. The definition is thus ethical from this view. But it is important to remember that, starting with Kant, there is a tradition stressing belief in universal human rights and human dignity as crucial for making a better world.

There is a paradox between the universalism in ethics and the particularity of law. Benhabib (2005) argues that these two have come closer to one another since the Second World War. She exemplifies with the founding of the ICC as a starting point, which has now led to an increased acceptance in questions of asylum and refugee rights in liberal democracies. This shows that it is not impossible to solve the paradox. Benhabib (2005) sets the starting point for this development after the Second World War. If she is right, the Nansen passports are a prehistory of cosmopolitan legislation. The ethical ideals behind them pushed them towards some kind of legislation, as when the individual states accepted them, and thereby the status of refugees.

The Nansen passports seem to advocate some kind of moral cosmopolitanism. For Nansen, in view of his earlier experience with the extremely poor conditions of the prisoners of war in Russia and of the starving population in Russia and Ukraine, it was not difficult to find philanthropic reasons for action to save these people. In his speeches to the Assembly of the League of Nations, Nansen often mentioned the necessity of rightful treatment of individuals who had become stateless as a consequence of inhuman wars and politics. While several members of the League considered humanitarian aid for Russia as support for a particularistic view, where human aid meant supporting a political system, maybe even to the disadvantage of others, Nansen kept arguing from his universalist standpoint. In order to solve the situation, it was necessary to see the humanitarian perspective: from a cosmopolitan point of view, see that these starving people were human before they were anything else, like political citizens. Putting the recognition of people as human beings before political realities could be understood as putting cosmopolitan understandings first. So, at least in a weaker meaning, the Nansen passports are premised on a cosmopolitan understanding. A cosmopolitan is a world citizen. But what would a world citizenship look like in this case? The difference between world citizen and citizen of a nation-state is that world citizenship is rather ethical, whereas national citizenship is state bound, and thus a legal condition.

To conclude, the earlier discussion has indicated that the Nansen passport could be considered as one of the very first serious efforts to construct a cosmopolitan citizenship. At the same time, it has been claimed that cosmopolitan citizenship can also be understood as a second-rate citizenship. In her gloomy writings about Europe after the First World War, the philosopher Hanna Arendt describes the refugees of the inter-war period as homeless, stateless and rightless:

> […] migrations of groups, who, unlike their happier predecessors in the religious wars, were welcomed nowhere and could be assimilated nowhere. Once they had left their homeland they remained homeless, once they had left their state they became stateless; once they had been deprived of their human rights they were rightless, the scum of the earth.
>
> (Arendt 1976: 267)

Is it possible that cosmopolitanism tries to offer a solution to each of these three problems or is it just part of the problem? The legal protection of refugees was from the very beginning a major part of solving the problem surrounding the stateless status of refugees. According to Arendt, the restoration of human rights can only occur through the establishment of national rights. One attempt to solve this kind of problem could be the post-national citizenship as defined by Soysal (1994), where she demonstrates how a national citizenship is unbalanced with the globally sanctioned norms and components of a supranational discourse where universal human rights have become individual rights, but where these are no longer protected by a traditional citizenship, not least because of global migration.

Conclusion

Can the almost forgotten Nansen passports be seen as cosmopolitanism in practice? Cosmopolitanism can mean different things, and it is important to discuss the concept from a historical perspective. In contemporary debates on cosmopolitanism three distinguishable fields show that the concept has multiple meanings. Nansen passports should be viewed in terms of cultural cosmopolitanism. Instead, they have some features that could be viewed from the perspective of moral and political cosmopolitanism, although political cosmopolitanism has more to do with the political structures and organization of societies. The Nansen passports were legally valid documents issued by the states that had accepted the League of Nations agreement, with support from the Nansen office. But as refugees had to re-apply for these passports annually, in the state of their current residence, these passports also had strong national elements, although they provided international protection, beyond national boundaries, which corresponds to cosmopolitanism in practice. The idea of moral cosmopolitanism can easily be found in the idea behind the Nansen passports. Nansen himself was often described as a person who stood 'above politics,' with his genuine arguing for decent conditions, especially for persons who had been struck by war, famine, disease and poverty.

With historical sensitivity one might ask if it is correct to call Nansen passports cosmopolitan in a time when 'cosmopolitan' was a bad word. The answer to that is yes, as the concept contains a practice and a view that prevailed also in the inter-war period. A larger problem is, however, that the Nansen passport did not imply civil rights. Holders of this document were allowed to travel, to work and to live legally in a country. But the holders also became second-rate citizens, as they did not carry the same rights as ordinary citizens. The passport was not universal. It was issued just to certain groups of refugees, despite the fact that many other groups also had lost their citizenship. The project, as it might be called, did not go as far as to include the refugees from empires that had been broken down by the new map of Europe. The immense number of stateless individuals faced great legal, social and cultural difficulties. They could not receive much help anywhere, as there was no consulate that could take responsibility for them.

Still, the conclusion is that the Nansen passport was a kind of preliminary cosmopolitanism in practice. The idea could be further refined, and in a way it has, as today's refugee politics is a direct development of the Nansen passport. The idea of the Nansen passports was indeed very constructive and creative. But because of its pioneering character it was difficult to oversee its consequences, as well as to see how it should be organized and administrated in the most suitable and fair way. The mission is still to be completed.

Notes

1 After homicide, war, famine and epidemics in the 1920s, the situation of Armenians was horrifying. Many of them ended their refugee status by emigrating to the Soviet Union. Approximately 100,000 Armenians went to Western Europe, and about half of them to France (Marrus 1985: 74–81).

2 Kwame Anthony Appiah has in fact defended the standpoint that human beings are responsible to one another and that there is an obligation to take care of strangers. Several communitarians have criticized cosmopolitanism, claiming that the responsibility to locals and compatriots is so strong that it makes cosmopolitanism impossible.

2 Between long-distance nationalism and 'rooted' cosmopolitanism?

Armenian-American engagement with their homeland

Tsypylma Darieva

> We get out of the plane… a smile runs across my face as I see Armenian writing and hear airport employees conversing in Armenian. Wait is it Armenian? It sounds like it, but I don't understand most of it. Oh no, my first feeling of culture shock. I get to the gate, fill out the paperwork and go straight to the immigration officer. I end up conversing with him for 10 minutes! He looks through my passport and asks me the most thought provoking yet simple question: 'what has taken you so long to visit Armenia?' Indeed, why has it? I had vacation time, I had the money, and I have the stamina to survive a long flight, so why not?
>
> (AVC Volunteer, spring 2007)

This interview excerpt describes the emotional experiences of one third-generation Armenian-American when he embarked on his diasporic homeland trip. It draws attention to a growing intensity of transnational interactions between members of second- and third-generation Armenian-Americans and the Republic of Armenia. In May 2007, during my fieldwork in Armenia,[1] I interviewed 20 young English-speaking volunteers of ethnic Armenian background who had travelled to Armenia as members of the Armenian Volunteer Corps (AVC): part of a 'three month programme to move mountains'.[2] The AVC, an international non-profit organization founded in 2001, 'calls on diasporic Armenians to volunteer their time, knowledge, and energy by living and working in Armenia to invest in the development of the homeland' and enjoy a meaningful exchange. There are many different activities included in this meaningful exchange programme: for example working in Yerevan's public organizations, schools, hospitals, non-governmental organizations (NGOs), helping to develop an impoverished village or to rebuild a church. Among them, one specific activity which attracted many volunteers was to plant trees in urban parks, neighbourhoods, tree nurseries and in the city of Yerevan and its suburbs, organized by the diasporic Armenian Tree Project organization in Boston.

Usually, this type of transnational activity, which includes identity claims and attachment of diasporic people to a specific territory, has been interpreted as a feature of long-distance nationalism that was opposed to assimilation processes (Glick Schiller 2005). As long-distance nationalists, diasporic members and migrants feel commonality with their homeland in terms of origin, history and identity, with borders and geographical distance playing no role in the articulation of their identity claims. The cross-border identities of migrants are understood to generate forms of 'long-distance nationalism'. Mobility in this case is seen to renew and reinforce bounded identities and social relations (Appadurai 1991; Calhoun 2002; Glick Schiller and Fouron 2001). This chapter is concerned with the possibility that these transnational projects can be inspired by, or produce, multiple, overlapping identities or cosmopolitan aspirations and projects. To date, such a concern has rarely been addressed. It is only recently that the mobilities of disempowered people and diasporic networks have been examined within the framework of cosmopolitanism (Glick Schiller *et al.* 2011; Skrbiš *et al.* 2004; Werbner 2005). The literature on transnational migration and diasporas has addressed the question of identity politics and its change, but it has not developed a theory of the cosmopolitical. Instead, most of this work has focused on political projects of migrant source states and migrants who live between their claims to a homeland and new land (Basch *et al.* 1994; Sheffer 2003).

This chapter attempts to re-conceptualize the nostalgic concept of homeland and, more precisely, the changing meaning of homecoming, a key feature of any diasporic identity. I examine the phenomenon of contemporary homeland trips or homecoming projects (symbolic or real) through the lens of social remittances, volunteering and philanthropy practices. Based on my ethnographic observations in the Boston area (MA, USA) and Armenia, the aim of this chapter is to explore a new diasporic rhetoric and transnational activities between Western countries, mostly the USA, and the Republic of Armenia.[3] In this context, the focus is on the diasporic Armenian transnational capacity to build on their cultural rootedness in order to develop shared practices of openness and local incorporation. My intention is to look at homecoming projects and diasporic engagement with the homeland beyond the ethnic lens (Glick Schiller *et al.* 2006) and to encompass diasporic travelling and sociability by the term 'rooted' cosmopolitanism. The term 'rooted cosmopolitanism', coined by Appiah, includes ideas and practices of transcending boundaries by relying on openness, inclusiveness and personal autonomy without giving up attachment to cultural and ethnic pasts and rooted identity (Appiah 2006; Werbner 2005).

These aspects of interconnectedness that link rootedness with openness cannot be seen in oppositional terms but rather as part of the creativity though which young diasporic travellers respond to global issues, like the environment and globalist discourses on human rights. Rootedness to territory and culture are assumed to be anti-ethical to openness, to the mutuality of shared human aspirations and sensibilities (Malkki 1997). This chapter challenges both these assumptions. By referring to this phenomenon as diasporic cosmopolitanism, I argue that this form of transnational engagement can exist together with particular diasporic ethnic ties. Noting that the growing literature on the 'cosmopolitan turn' emphasizes the plurality of cosmopolitanisms (Glick Schiller *et al.* 2011; Nowicka and Rovisco 2009; Rapport and Stade 2010; Robbins 1998a,b; Vertovec 2009; Werbner 2008b), this contribution offers insights into the way networks of interconnection and exchange between a new homeland and the diaspora are reconsidered by some members of the second and third diasporic generations themselves. I suggest that a 'cosmopolitan dimension' and the maintenance of ethnic ties occur simultaneously in the way the US American-Armenians express their motivations and define the flow of their social capital and donations to the so-called diasporic homeland – the Republic of Armenia.

Long-distance nationalism among Armenians

The Armenian diaspora is considered to be a paradigmatic diasporic group associated with strong affiliation to ethnic roots, forging long-distance nationalism and political ethnocentrism reified in the notion of an ancient culture that has been recently reshaped by a strong institutionalized identity of victimization. Much has been written about Armenian diasporic nostalgia, its diverse representations and manifestations of cultural memory concerning the Armenian massacre in 1915 and the ways this memory forms diasporic belonging across borders and generations that has not disappeared to this day. It is not new to draw attention to the variations in intensity and goals of the ways in which diasporic members can combine transnational 'rooted' and assimilative strategies. As noted by Peggy Levitt (Levitt and Waters 2005), it reveals that transnational ethnic practices and assimilation are not diametrically opposed to each other. Instead, as this chapter shows, transnational ethnic practices and assimilation create diverse interconnectedness across different generations.

My main argument is that travelling and 'homecoming projects' have increasingly less to do with a national 'nostalgic' re-enactment of the past and vengeance for the past, but rather with a 'journey to the future' and a creative and to some extent naïve ideal of a cosmopolitan '*Weltverbesserer*' (world improver). This practice should be analysed more carefully and differentiated. It seems that this kind of homeland trip is more likely to be relevant for members of the third generation of Armenian-Americans. Members of the second generation, like their grandchildren, are interested in engagement with a homeland, but in a specific way. According to my interviews and observations, members of the second generation interact with Armenia transnationally without leaving the territory of their new homeland in either the USA or Canada, but simply by supporting their children financially and by giving and donating money and social energy to non-profit organizations and community centres. They are very heterogeneous with regard to socio-political characteristics and their intensity of involvement in the activities of ethnic Armenian organizations. This chapter deals only with 'activists' and networks at the individual and institutional levels which are visible in transnational social fields. One feature that many American-Armenians share is their social status in the USA, as they belong predominantly to the middle class.

By transferring social and economic capital to help a poor country, Armenian-Americans gain a feeling of incorporation into their separate 'sacred homeland', but also the global issues of 'development' and 'democracy'. This new energy and moral idealism enable diasporic people to rethink the cohesion and existing boundaries between 'us' (the diaspora) and 'them' (the homeland). Today, they reclaim Armenian soil through activities that contribute to the global issues of human rights and the environment, issues that affect the entire planet and its inhabitants.

Diasporic homecoming between long-distance nationalism and cosmopolitan projects

The meaning of diaspora experienced a significant shift and moved away from its identification as a bounded localized minority community in a destination country or simply a cultural formation in a globalized world. For a long time, the term 'diaspora' had been exclusively used for defining three groups: Jewish, Greek and Armenian diasporas (Clifford 1994: 305; Cohen 1997: 507–508). At the beginning of the 1990s, the discussion on diasporas began to include many other cases and, as Tölölyan (1996: 4) argued, the term 'diaspora' 'now shares meanings with a larger semantic domain that includes words like immigrant, expatriate, refugee, guest-worker, exile community, overseas community, ethnic community and minority', even when they have been largely assimilated. This rather broad conception has led to substantial criticism of the term 'diaspora' (Brubaker 2005). Thus, today the term 'diaspora' is associated with diversified transnational movements of people between exodus and destination countries, with mobility of capital, commodities and cultural iconographies (Brah 1996). In the 1990s, diasporas became mobile and powerful, resulting in a growing diversity of alternative identities. Werbner (2008b) noted that diasporas are characterized by both ethnic-parochial and cosmopolitan stances.

Brubaker (2005) criticized the treatment of diasporas as unitary actors (*the* Kurds, *the* Tamils, *the* Russians). Similar to Clifford (1994), he proposed that diaspora should be perceived as a claim, idiom or stance, as a way of 'formulating the identities and loyalties of a population' (ibid.: 12). Finally, in order to avoid the reification of diasporas, Brubaker (2005) suggested avoiding the noun 'diaspora' and using the adjective 'diasporic' instead, so as to speak of diasporic projects, diasporic claims, diasporic religion and so forth (ibid.: 13). Following Brubaker's suggestion to 'de-substantialize the diaspora', this chapter treats the term 'diaspora' as a category of practice, claim and imagination, rather than a bounded group. Though migration and global diaspora movements only recently became a significant part of anthropological interest, scholars have moved away from conceptualizing social relations and identities as bounded, whether they have been envisioned as impossible and territorially rooted cultural differences or by the national borders of nation-states.

In contradistinction to the 'de-substantializing of diasporas' in the Western world, the collapse of the Soviet Union brought with it a counter process of emerging new diasporas within its former boundaries and their massive proliferation in politics, media and everyday life. As a result, the use of the term 'diaspora' has penetrated the academic, political and public discourse in Russia and the former union republics. For example, in 2010, a search of the Russian internet (RuNet) for the word 'diaspora' (*диаспора*) using Google yielded 1,110 hits. Another popular Russian search engine, www.yandex.ru gave more than two million hits for 'Armenian diaspora'.[4] Accordingly, it can be assumed that homecomings and return migrations, which previously were referred to as 'silent migration' (Stefansson 2004), seem to be increasingly significant for a 'voiced' global mobility, including post-socialist societies in Eurasia. Noteworthy here is that the rise of 'national' politics in the East and West also plays an important role in creating spaces for migration based on primordial ideals of ethnicity and 'the myth of home'. Although 'homecoming' was unimaginable for many people before the demise of the Soviet Union because of closed borders, now temporary or permanent 'homecoming' is possible. Acknowledging this change in circumstances, Tishkov (2003: 467) and Kosmarskaya (2006) even speak of a 'passion for diaspora' (*diasporalnoe pirshestvo*) in Russia and an ongoing process of 'diasporization of the whole country'. As result both argue that the term diaspora should be handled carefully, even rejecting its use as an analytical tool.

One of the key features identifying members of a 'diaspora' is their continuing attachment to the homeland, regardless of whether it is an imagined or real country of exodus. Much has been written about the ideals and paradigms of the diasporic identity, but there have been fewer investigations of the ways diasporic people practice this kind of attachment in the transnational age. Attachment to the homeland can take many different forms and meanings. It can be expressed in the construction of an imagined community with a sacred place reserved for worshipping the land of exodus. It can also be expressed in the activities of political associations with territorial claims, in repatriation movements, in artistic expressions of nostalgic longing for home or in simply hanging an image of the homeland in the living room. But in addition, the search for roots can drive financial investment and temporal mobility among the second and third generations. This form of homecoming becomes popular among some of the younger members of the third generation of Armenian- Americans, as the subsequent discussion will show. From a social anthropological view, homecomings are seen as an emotive moment in a migrant's life cycle based on ethnic and cultural ties to their ancestors (Basu 2005; Holsey 2004; Stefansson 2004). This emotional aspect can be traced in the statement of the AVC member cited at the beginning of this chapter. According to this definition, it is expected that a delocalized religious or ethnic minority at one point in time 'goes home' to the former community or nation-state from which it migrated. An individual decides to return to his or her birthplace, to watch the sunset and then be buried in their native soil.

On the global level, return has become part of a broader discourse on human rights and property restitution, intimately tied to the legitimization of regime change and state-building in post-authoritarian countries. Some scholars define return and homecoming movements as non-progressive, more as a possibility of 'de-globalization' (Olwig 2004), the end of migration and even as a counter-movement with regards transnational paradigms.

Current trends suggest that the classical form of homecoming as return migration and repatriation is losing its ability to attract second- and third-generation diasporic people (Levitt and Waters 2002). But it does not mean that the third generation is entirely assimilated. Instead, new ways of engaging with the homeland within global social movements are emerging and seem to play an increasing role in the reconfiguration of relationships between diaspora, the host land and the homeland. In this chapter, I give some insights into this form of transnational engagement by differentiating the scope and meaning of the engagement with the homeland between second- and third-generation Armenian-Americans. More specifically, I refer to the practice of volunteering and tree planting in Armenia sponsored by diasporic non-profit organizations.

Homeland trip motivations: the third generation

Armenians who comprise the diaspora call themselves *Spiurk* and Armenians from the Republic of Armenia are known as *Hayastantsy*. These two notions characterize differences along political, social and cultural lines. Historians identify *Hayastantsy* as Transcaucasian or Russian Armenians and *Spiurk* as Ottoman or Turkish Armenians due to the geographical and political divisions between two empires in the nineteenth century. A significant number of diasporic Armenians were expelled from the former Ottoman Empire (today Turkey) and not from the territory of modern-day Armenia.[5] They settled in the USA at the beginning of the twentieth century not only as refugees but also as labour migrants. At the end of the 1980s, Armenian-Americans were characterized by Anny Bakalian (1994) as 'feeling' and not 'being' Armenian anymore. Despite a high level of assimilation and social mobility, there is still a remarkable level of institutional completeness in terms of community organizations (schools, churches, media, museums, charity organizations) and political lobbying in the USA.[6]

The number of contemporary diasporic newcomers in Armenia is not high, but it is generally believed that they have a significant political and economic impact in Armenia. After the collapse of the socialist system, Armenia suffered tremendously from the subsequent social and economic transformations as well as the violent conflict between Armenia and Azerbaijan over Nagorny Karabakh. As a result, in the early 1990s, Armenia experienced a high rate of emigration to Russia and other countries. Whereas, today, Armenia's neighbouring countries, such as Georgia and Azerbaijan, are considered to have been shaped by internal forces such as the effects of the Rose Revolution and oil businesses, respectively, Armenia, as an impoverished and isolated country became a recipient of labour migrant remittances and of a large amount of donations and know-how input from diasporic networks in Western countries and international organizations. Although the investment and remittances have not significantly reduced the level of poverty in Armenia, Armenia continues to be more dependent on remittances and international aid than other post-Soviet, South Caucasian societies.[7]

Noteworthy here is that the migration of diasporic Western Armenians from different parts of the world to Armenia is not a new phenomenon. After the Second World War, between the 1940s and 1960s, Western Armenians enthusiastically participated in the state repatriation programme known as *nerghakht*. Attracted by Stalin's campaign to repopulate the regions of Kars and Ardahan, which were contested with Turkey, about 100,000 Armenians from different countries resettled in Soviet Armenia. A formal territorial claim was made by the Kremlin to the Turkish ambassador in Moscow, but it was dropped in 1949 with no border change. The dramatic experience of the return programme (including Stalin's deportation of newly arrived Armenians to Siberia for being 'bourgeois elements' and the precarious living conditions and unemployment) disillusioned diasporic Armenians in many parts of the world and created political tension between the diaspora and Soviet Armenia (Panossian 2006; Suny 1993; Ter-Minassian 2008).

A unique feature of the Armenian homecoming in the twentieth century is that instead of returning to the actual ancestral places, the hometowns and villages in the Eastern Anatolian plateau (Turkey), grandsons and granddaughters instead invest, engage and settle in Armenia and the land neighbouring Turkey.[8] Therefore, Armenian-American diasporic visitors have no local dimension and intimate knowledge of a particular genealogy, place, or village in the territory of Armenia. It is not surprising that, unlike so many migratory transnational networks built on a foundation of individual informal ties of kinship and remittances to family members and an obligation to support local households (Armenians in Russia, Indians, Chinese or Ghanaians in the USA and Europe), members of the US-Armenian diaspora build homeland ties primarily through formal NGOs and international organizations. This tendency by some diasporic groups to invest in countries other than their 'source countries' (such as Croats from Serbia who invest in the newly independent Croatia) has rarely been studied by migration scholars. The question is whether this specific trajectory of travel influences the scope and intensity of engagement with the homeland and the tendency to combine ethnic parochial claims with globalized and universalist values.

According to my observations and interviews, there is a shift in intensity in the way Armenian-Americans engage with the Republic of Armenia. In the early 1990s, some second-generation Armenian-Americans were interested in long-distance 'cultural' and individual engagement, like symbolic re-burying family members in Armenian soil, bringing family relics to local museums in Yerevan (Darieva 2006). Today these Armenians donate and invest in infrastructure projects like roads, the greening of urban parks, housing for refugees, poverty reduction in villages, and environmental projects. This kind of transnational relationship can be recognized as social remittance, where migrants and diasporic people transfer resources, ideas and behaviour from receiving to sending countries (Levitt 1998, 2002) and remain oriented to the communities they come from. However, in the case of the Armenian diaspora in the USA we are dealing with a specific pattern of social or collective remittances, which are oriented not to their actual ancestral territory in Turkey, but to the neighbouring country.

My recent observations and interviews reveal that many second-generation Armenian-Americans have never visited Eastern Anatolia or Armenia; instead, they interact with the 'homeland' transnationally without leaving their environment, by investing money and supporting their children's mobility. Consequently, the third-generation American Armenian youth is increasingly engaged in face-to-face participation in transnational social fields. Regarding their motivations, diasporic people arrive in Yerevan not just to see the holy Mount Ararat (*Agri Dagi*), but rather to 'develop Armenia'. Diasporic travellers talk about the transfer of ideas, cross-cultural exchanges of material, and know-how to a developing country, often making reference to their broader global aspirations. As Ishkanyan (2008) noted, the myth of return and patriotism in the twenty-first century appears to be weakening, as most diaspora Armenians prefer to interact with Armenia transnationally and not as a one-way process.

There are dozens of visible, larger non-profit organizations working in the education and health sectors in Armenia. Among many diasporic profit and non-profit organizations and NGOs, there are only two target oriented organizations offering 'homecoming' support for members with an Armenian background: the AVC and *Birthright Armenia*, similar to Jewish and Greek diasporic youth organizations. Founded in 2001, both organizations are engaged in transnational activity to support volunteers in Armenia, including those who grew up in Western countries but who have at least one Armenian grandparent. Between 2007 and 2009, more than 200 male and female volunteers from the USA, Canada, France and Australia between the ages of 21 and 34 went to Armenia as a part of the AVC.[9] The number is growing. Without nationalistic slogans, its goal is empowerment and a desire to join others from around the world working to save the planet. Politically, AVC statements differ significantly from the goals of nationalist diasporic Armenians (*Dashnaks*), who until recently wanted to annex lands in Eastern Anatolia inside Turkey and to establish an Armenian state (Phillips 1989). The AVC recruits young volunteers through a humanitarian rhetoric and focuses its efforts on the territory of the Republic of Armenia. Explaining his drive to settle in Armenia for one or two years, one 30-year-old male volunteer from Boston emphasized: 'There are lots of things to change here. You know, there is a problem with poverty, infrastructure. There is a problem with corruption' (Yerevan, 7 May 2005).

These temporary visits, with a duration of three months to two years, are often described by these young volunteers as a kind of philanthropic 'giving' or individual adventure, as well as a symbolic part of a 'homecoming project'. In both cases, the motivations and aspirations of volunteers and of the institutions that mediated their travel have been framed in discourses of a 'meaningful exchange' of skills and know-how between developing and developed societies. Conducting a group interview with eight young volunteers in Yerevan on 5 December 2007, I posed a question regarding the direct and indirect motivation to participate in this programme.[10] All the respondents stated that their travel was not necessarily related to a 'natural' behaviour of diasporic descendants. A volunteer from Australia explicitly emphasized the individual and pragmatic dimension of her goal:

> I came here primarily as a volunteer…it was not so much about Armenia as it was about me coming here to help people, as I had no real links with my homeland before I came… My travel is kind of giving me a big kick, showing me where I want to be…

Another informant stressed his aspiration of engaging with Armenia as a 'perfect time' for a life-stage event and as a good place for collecting life experience. Similar to the experiences of other European second-generation transnationalists (Turkish Germans, Swiss Italians, Greek Germans or British Pakistani) the 'homeland trip' was made in a quest for personal freedom and self-realization (King and Christou 2010; Wessendorf 2009). Victoria from Washington, DC, talked of her decision to travel to Armenia:

> I am not tied down with family, with a career, with all these things and it is a perfect, perfect time to travel. And I always knew that it was the place where I would want to go. There are a lot of life experiences that I think are necessary to gain right now and it is a great time to do that.

Australian-born third-generation Serena, who travelled to Armenia for nine months, talked of the adventurous side of her 'homeland trip':

> I am trying to make my everyday life like an adventure, because I am not in Australia now, and do not have to stick with this or that job, do not have any responsibilities.

Victoria from Washington, DC, simultaneously referred to the individual life-stage reason and to the role of parental obligations, in particular her parents' dream of visiting their cultural homeland:

> It was always a dream of my parents and grandparents, but it was never the right time. They would say that the situation was not good or safe enough, because of corruption or whatever. But after I came in 2000 and told them how amazing it was, how beautiful it was, everybody wanted to come to Armenia. They came here on holiday in 2003.

Finally, many interview respondents emphasized the cultural heterogeneity of the social environment in which they grew up. This aspect is related to the fact that at least four of the interview respondents identified themselves as being half-Armenian and having grown up in ethnically mixed families. Many of the volunteers explained that their motivations were not influenced by their parents' ambitions, but rather their decision to come to Armenia was independent of their parents. Moreover, the majority of the volunteers I talked to during my field work viewed their transnational activity and behaviour in a different way to their parents, the latter seemingly following a 'sedentary' pattern of interaction. Lucia (26 years old), whose father is of Armenian descent and mother Austrian, explained her experience in this way:

> My father has never been to Armenia. I was hoping that I would get him to come to Armenia during my stay, but unfortunately he is not going to come. I suppose he is scared to come to Armenia to see how ideal it is not.... It is not the utopian ideal society that he kind of wants it to be.

Only a few of the volunteers' parents had actually travelled to Armenia before their children visited the 'sacred land'. Thus, their transnational engagement with the homeland differs in character and intensity, but it does not mean that they do not participate in transnational social connectedness. As mentioned previously, some parents prefer to support transnational organizations by donating and investing money in their children's travel or in maintaining transnational projects.

Between metaphors of rootedness and cosmopolitanism: the second and third generation

One of the areas where diasporic notions of attachment to the homeland is pronounced in a less nationalist framework and rhetoric is that of the new form of social remittance which stresses the value of individual engagement, development and ecological harmony on a global scale. In the following section, I give some insights into this practice which was developed in Watertown (Boston area, MA) by the *Armenian Tree Project.*

In spring 2005, an English language website announced their intention to plant 1.5 million trees in memory of the victims of the Armenian genocide.[11] This effort was organized by the Armenian Tree Project (ATP), which has its headquarters in Watertown in the Boston area, and conducts environmental programmes in Armenia's impoverished and deforested zones. The ATP was founded in 1994 in Watertown and in Yerevan by Carolin Mugar, a second-generation Armenian-American, whose father left the village of Kharpet in Anatolia in 1906 to settle in Massachusetts. Later, between the 1940s and the 1960s, Mugar's father and his brother became very successful businessmen, establishing the popular US supermarket chain *Star Market*. The Watertown ATP office brings a large amount of capital into Yerevan, opening nurseries, planting trees and starting projects in surrounding villages.[12] The economic resources derive mostly from public fundraising ceremonies organized by the ATP in the Boston area and elsewhere among Armenian-Americans. These events are not huge transnational festivals or face-to-face collections, as is often the case for Indian and African fundraising rituals (Nieswand 2009), but are more like bureaucratized appeals for funding via the Internet and mailing lists.

The local office's activities in Yerevan are divided into three main tree-planting sites: community sites in the city, nurseries, and impoverished villages with a high percentage of Armenian refugees from Azerbaijan. The idea of planting trees began with the practical goal of preventing topsoil erosion and supporting fruit production among villagers after the end of the Karabakh conflict in 1994. Soon, the ATP turned to renewing urban parks and community tree planting and, finally, the Armenian Tree Project expanded its activities to larger projects such as reforestation and environmental education programmes in the Lori region, in the northern part of Armenia.[13]

Memory of the Armenian losses in 1915 has become a powerful symbol for successful fundraising campaigns within diasporic networks in Massachusetts. However, the ATP has not only received generous support from a cluster of US Armenian family foundations but also from international such as *Conservation International* and the *World Wildlife Fund.* Along with informal individual activities, international non-profit organizations play a significant role in these homecoming projects, which seems to shape the nature of philanthropic transfer and the relationship between the Armenian homeland and the Armenian diaspora. This raises the question of how these new relations alter the nature of political and ethnic claims among second-generation Armenian-Americans? To what extent does this kind of diaspora philanthropy and transnational engagement differ from previous humanitarian aid programmes implemented by different diasporic organizations, such as the AGBU?[14] It would be too ambitious to provide a detailed answer to these questions, but I will nonetheless attempt to approach them.

Tree planting in Armenia has had a particular transnational impact on life-cycle rituals (birthdays, anniversaries and deaths) within the Armenian diasporic organizations in the Boston area. Increasingly, diasporic people donate to the ATP in commemoration of a family member. Another transnational technique developed by the ATP is a 'Green Certificate', which can be presented to donors confirming their sponsorship of tree planting in Armenia. Increasingly, donors also make pilgrimages to the sites where sponsored trees were planted and to nurseries in Armenia.

The projection of the homeland as an evergreen landscape, which has been created by the ATP, is built on European and North American romanticized images of nature. In its aesthetic design and combination of colours, the tree landscape differs from the traditional representations of the Armenian garden, which uses vineyard metaphors. However, the romantic ideals and cultural practices of tree planting have parallels with the perception of the 'world as a garden', as rooted in the pre-Christian cultures of the Near East (Petrosyan 2001).

The tree-planting culture seems to have diversified the typical Armenian image of the homeland, which has been focused on Mount Ararat. The reproduction of the mountain, as depicted in various genres of landscape painting, is still omnipresent in Watertown community centre buildings. A 'rooted' evidence of the transnational revitalization of Armenia, based on the tree symbolism, was also created at Karin ATP Nursery and Education Centre. Fixed on the wall inside the centre is a depiction of the Tree of Life. One can see a metal tree of an unrecognizable type with many leaves. The leaves on the tree serve as small plaques for engraving numerous individual names of donors and volunteers, all written in the Latin script. On the left side of the wall, brass shafts of sunlight are fixed over the Tree of Life. The rays represent the unity of those who donate significant sums to the development of tree nurseries. However, the ATP's official logo design is three triangular green trees, which is similar to the design on oriental rugs. Flyers, websites, newsletters and donation certificates are identified by an image of three evergreen trees without any specific mountain images. Boththe mountain and the trees are essentializing symbols of nature. But unlike the mountain, which is highly associated with a particular historical longing for a past and a loss, a tree represents social qualities, such as vitality, cultural universality, and a powerful orientation towards the future.[15] In general, there have been strong connections between environmental romanticism and nationalism since the end of the nineteenth century (Lekan 2004). Thinking in terms of roots and trees reminds us of the ubiquitous yearning of nations in the search for their roots in an ethnic past (Malkki 1997; Smith 1986). However, unlike the Zionist forestation campaign in Israel (Bardenstein 1999; Braverman 2009), which pursued the aim of putting down roots in a new place and reclaiming the territory to the exclusion of others, tree-planting actions in Armenia serve as a social marker of a new transnational bond between the homeland and the diaspora, and between a small corner of the world and global issues.

In the Armenian case, national imaginary can stimulate a simultaneous inclusive globalism, so that universal and cosmopolitan ideas and practices are implemented within redefinitions of homecomings and transnational narratives of reconnecting the diaspora with the homeland. The rhetoric of the Armenian Tree Project tries to create a new dimension for envisioning a mutually acceptable future that diminishes the tensions between 'us – *Spiurk*' and 'them – *Hayastantsy*' via global issues. In 1998, for example, the ATP jointly initiated an event to mark Earth Day and Arbor Day in Armenian villages.[16] The date, 22 April, coincides with Vladimir Lenin's birthday, the traditional day for celebrating volunteer work initiated by the Soviet authorities. In the Soviet period, everyone (Soviet institutions, schools, enterprises, etc.) was required to mark this day by cleaning the immediate area around which they worked or studied and then plant a tree. This day has since been transformed into global Earth Day. As reported in the ATP newsletter in 2007, the celebration at an ATP nursery in the village of Karin '…united Armenian officials, ambassadors, NGO representatives and the local population in a tree-planting ceremony to raise awareness of ecological issues and emphasize the need to solve them together…'. The director of the Armenian Tree Project, Jeff Masarjian, stated that the environmental movement had become a global phenomenon by the early 1990s, when 200 million people from around the world started celebrating Earth Day.

In an ATP newsletter from spring 2007, one can read the official, twofold vision and pledge to Armenia: '*We will use trees to improve the standard of living of Armenians and to protect the global environment*' (2007: 2). This quotation indicates that planting trees simultaneously brings to mind a naturalized and ethnicized connotation based on the typical diasporic search for roots of renewal, and is also re-conceptualized within broader global frameworks. By positioning actions within a movement to sustain and protect the planet, the act of tree planting to help Armenia is transformed into a form of creative cosmopolitan discourse. The newsletter also states:

> We are proud to join the international effort to plant trees to fight climate change, which is worsened by rampant deforestation around the world. In 2006, the ATP joined the worldwide tree planting campaign launched by the 'Billion Tree Campaign'.
>
> (2007: 2)

Whereas the Zionist project is characterized by a mono-cultural use of the pine tree (and a physical occupation of the land through planting pine trees), promoting an ethnically driven security agenda (Braverman 2009), the Armenian Tree Project, in both donation and landscape greening techniques, is not fixed to the ecological symbolism of any particular tree. Instead, it emphasizes Armenia's biodiversity in a global context and sees Armenia as part of a larger region – the Caucasus.

Having started with planting fruit trees for villagers, the ATP now plants a wide range of decorative trees as well as forest trees: North American and Eastern Asian thuja, Crimean wild rose, Chinese magnolia, etc. As a part of an international project, ATP tree planting is linked to a commitment to biodiversity, which is made explicit in the curriculum for environmental education published in English and Armenian:

> There is biodiversity within a forest. Forests contain many communities that support diverse populations of organisms. Different forests have different levels of biodiversity. Armenia has a complex relief, as a result of which the regions have strongly differing natural climatic conditions (e.g., precipitation, temperature, topography, etc.). These variations lead to different forest communities with differing species, thereby contributing to Caucasian biodiversity. Armenia is considered part of the world's 25 most ecologically diverse ecosystems by the World Wild Fund for Nature.
>
> (Wesley 2010: 16)

Conclusion

Contemporary Armenian programmes challenge the ethnic idea of homecoming through cosmopolitan rhetoric and discourses framed as 'progress', 'democracy' and 'global civic society': i.e. the current Armenian homecomings not only comprise anti-modern, de-globalized return migration (Olwig 2004) but also modern long- and short-term visits, work contracts, development-aid programmes, and social projects across borders. By transferring social and cultural capital to a poor country, the US Armenian-Americans gain a feeling of incorporation into the separate 'sacred homeland' as well as into the global issues of 'development' and 'democracy'. This observation leads to the question: Since a significant number of Armenian-Americans give donations to repair roads and hospitals, museums and churches in Armenia, to what extent do these gifts and diasporic philanthropy contribute to a notion of transnational or cosmopolitan morality? Re-evaluations of diaspora lifestyles across different generations do not provide a set of answers on the quality of cosmopolitanism, but do provide a way to distinguish different attitudes and ways of engaging with the homeland.

On the one hand, the centrality of planting trees in one particular place – on Armenian soil – is based on ethnic notions of symbolic repossession of the lost land. On the other hand, trees and their symbolism are associated with a progressive ideal of a shared global future for the inhabitants of the planet. In that sense, appropriating a language of global ideals about the 'green planet' helps to find new forms of solidarity and sociability that move beyond the ethnic desire to restore a greater 'historical Armenia' and return to the place of origin. I define cosmopolitan sociability as forms of competence and communication skills that are based on the human capacity to create social relations of inclusiveness and openness to the world. Considering the concept and the values of 'bifocality', I identify a diasporic cosmopolitanism, where rootedness and openness cannot be seen in oppositional terms but constitute aspects of creativity through which migrants build homes, secure their existence and legitimize their diasporic status within transnational networks. These competencies take many forms and open an array of possibilities, and some of these possibilities emerge as cosmopolitan.

Cosmopolitanism is often equated with the experience of mobility in the modern world. I am against making teleological connections among mobility, transnational networks and cosmopolitanism. In this sense, it would be wrong to assume that diasporic people and the nature of transnational networks develop evolutionally from ethnic/national members of the first generation, to transnational members of second-generation Armenian-Americans and finally to cosmopolitan members of the third generation. It is rather an overlapping process, with a temporal modus and embedded in different local contexts.

Again, there is a tendency to compare Armenian diasporic experiences with the Jewish case, but the Armenian engagement with the homeland should not be equated with the Jewish Zionist movement. Armenian-American diasporic visitors have no local dimension and no intimate knowledge of a particular genealogy, place, or village in Armenia. The homeland attachment is directed to a territory that is not the ancestral homeland – the territory of Turkey from which their grandparents actually originated. Therefore, it is not surprising that unlike so many migratory transnational networks that are built on a foundation of individual, informal kinship ties and an obligation to support family members (Indians, Chinese, Ghanaians), members of the US American-Armenian diaspora build homeland ties primarily through formal NGOs and international organizations and thus without a feeling of indebtedness to the symbolic homeland.

Many of the third-generation diasporic Armenians combine 'homeland imaginaries' and 'ancestral tourism' with an assertion that to reclaim Armenian soil is to contribute to development of the entire planet and its inhabitants in terms of environmental issues, gender relations and civil society.

The ties between the homeland and the diaspora are relatively weak and the diaspora's support for Armenia is less institutionalized and less 'strategic', but more individualistic and project-specific. Although the imaginaries of home and practices of a diasporic 'trip to the homeland' are framed in terms of remembering ancestral origins and traumatic loss, these trips take on new dimensions and are often framed as individual projects.

Given the significant cultural, socio-political and historical diversity of homecoming forms and the different ways of dealing with mobility, to talk of common characteristics of a culture of diasporic homecoming and transnational engagement is sometimes misleading. In my mind, the anthropological approach to understanding diasporic cosmopolitanism and diasporic activism is not only to analyze the ethnic roots and social order within which migration takes place but also to study how social change is affected by migration, to study the transformations in both host and homeland societies and generational change.

Notes

1 The ethnographic field work was conducted within the research project 'Identity Politics in Societies in Transition. Armenia on the Way to Europe?' (2004–2008) at Humboldt University, Berlin and funded by the German Research Society.
2 See http://www.armenianvolunteer.org (accessed 11 February 2012).
3 Observations are based on ethnographic fieldwork conducted between April 2005 and December 2007 in Yerevan, Armenia and Watertown, USA, a suburb of Boston. Following the approach of multi-sited ethnography, my 'field' encompassed geographical sites between Armenia and the USA (Yerevan and Boston, Watertown). For my research, I conducted interviews with activists of diasporic non-profit organizations and donors in the Boston area, employees of these organizations in Watertown and Yerevan, as well as with local Yerevan city dwellers.
4 Accessed 26 January 2012.
5 I leave Iranian Armenians aside.
6 The community activists I interviewed during my fieldwork in the Boston area told me that in the 1980s, some American-Armenians started to 'return' to the typical Armenian name-ending '-yan' in their family names: for example, from Toovmajan to Tomasson and from Tomasson back to Toovmajan. Many second-generation Armenian-Americans received European names from their parents and until the end of the 1960s it was not unusual to cut the Armenian ending or to modify names towards the dominant Protestant names.
7 See Pearce, K.E. (2011) 'Poverty in the South Caucasus', *Caucasus Analytical Digest*, No. 34, 21 December.
8 A hostile attitude towards Turkey among the Spuirk still keeps Armenian-Americans from visiting Turkey.
9 Interview with the AVC officer was conducted on 16 May 2008.
10 The average age of male and female volunteers was around 24–27 years and they came from different Western countries (USA, Australia and Austria). Beyond working in the spheres of education, health, community development, social projects,

IT and medicine, they actively participated in public life, which allowed them to engage in visible, ongoing social projects supported by AVC. For example, the majority of the interview respondents took part in a march against domestic violence in Armenia, organized by the Women's Resource Centre, Boston in cooperation with AIWA (Armenian International Woman Association).

11 See http://www.armeniatree.org (accessed 22 January 2012).

12 Although only three employees are working in the Watertown ATP office, more than 40 local employees were hired in Yerevan.

13 According to the ATP's own estimation, the survival rate for the trees planted is 86 per cent, once they make it past three years – a very high survival rate.

14 AGBU, The Armenian General Benevolent Union, is a non-profit organization established in 1906 whose mission is to preserve and promote Armenian identity and heritage through educational, cultural and humanitarian programmes. For more details, see Melkonyan (2010).

15 See Rival (1998) for more depictions of the symbolic significance of trees in a number of cultures.

16 Introduced in the middle of the nineteenth century, Arbor Day was proclaimed by President Richard Nixon as National Arbor Day in 1970 and was fixed on the last Friday in April. Arbor Day is also now celebrated in other countries. Variations of this day are celebrated as 'Greening Week' in Japan and 'The New Year's Days of Trees' in Israel. For a more detailed critical study on history and political meaning of planting trees actions in the USA, see Cohen (2004).

Part II

Making and re-making diasporas from former Yugoslavia

3 Diaspora impact on European Community policy-making

Ex-Yugoslavia as a case study

Branislav Radeljić

Diaspora populations are capable of penetrating developments of both international and domestic politics. While being free to develop particular policies independently of their representative authorities, '[t]hey may support or oppose the government of their home country, morally, financially, and as suppliers of weapons and even personnel to the faction they favor' (Esman 2009: 8). This statement is a clear indication of how serious diaspora activism can be: it is their position, between homelands and adopted countries, which provides them with noticeable power with regard to policy-making. Moreover, this long-distance ability of migrants is often characterized by sets of irresponsible actions – a circumstance where none of the 'shrewd political manipulators' is likely to bear responsibility for the actions undertaken (Anderson 1992: 13).

This chapter elaborates on diaspora activism during the Yugoslav crisis of the early 1990s by focusing on the role played primarily by the Slovenian and, to a lesser extent, Croatian diaspora communities. This primacy is justified by my argument that the Slovenian diaspora lobbied for the recognition of both Slovenian and Croatian independence by the European Community (EC). While the Brussels officials were trying to find a solution for the Yugoslav drama, diaspora groups had already established and worked on their common objective. In fact, their activism contributed to the overall EC policy – a policy that could be fully understood by looking at a new cosmopolitan Europe. Cosmopolitanism implies an emphasis on appreciation of difference and universal inclusion, two aspects promoted by elites that take on an increasingly important role in times of crisis (Beck and Grande 2008; Rumford 2007). Indeed, in the case of the former Yugoslavia, the Europeans expressed solidarity and acknowledged the transnational aspect of the cosmopolitan approach. They listened to what the Slovenian and Croatian diaspora groups were saying about the Yugoslav federation and approved what the groups desired as the outcome of the state crisis: international recognition of their homelands. For the clarity of arguments presented, some parts of the chapter examine the two diasporas separately; however, often, and as will be particularly obvious in the section discussing their activism and power at the EC level, the two are presented as a unitary actor.

Two diasporas, one objective

The Republic of Slovenia felt frustrated within the Yugoslav Federation. Its growing nationalism was a result of economic hardship and political divisions that Yugoslavia was facing throughout the 1980s. One scholar described the position of his fellow Slovenes as '[b]eing deprived of their enjoyment by "Southerners" (Serbians, Bosnians) because of their proverbial laziness, Balkan corruption, dirty and noisy enjoyment, and because they demand[ed] bottomless economic support, stealing from Slovenes their precious accumulation by means of which Slovenia could already have caught up with Western Europe' (Žižek 1990: 55). The discourse about Slovenian economic advancement was easily complemented with political and ideological dimensions that, taken together, made the quest for independence even stronger. With regard to the members of the Slovenian diaspora, they perceived the end of the communist era 'as an opportunity to materialize their political agenda' (Skrbiš 2007a: 149).

Accordingly, the Slovenes began establishing new diaspora organizations and contacts within internationally recognized institutions. As Zlatko Skrbiš (2002: 47–8) put it:

> The processes which led to the proclamation of Slovene independence from Yugoslavia in 1991 were based on the idea of global Slovene solidarity, aimed at transcendence of historically conditioned political particularisms. This idea of solidarity conditioned the establishment of the Slovene World Congress, an institution designed to foster a sense of an overarching commitment to fostering a Slovene national community as a transnational project.

In fact, the Congress represented a direct linkage between the homeland and diaspora. It promoted Slovenian interests around the world and, more relevantly for us here, it managed to strengthen the relevance of the Slovenian voice within European Community leadership.

With regard to the Croatian diaspora, it was during the 'escalatory stage' of the Yugoslav state crisis (1987–1991) when the Croatian diaspora began to rely on the ethnic tensions in their homeland and illusionist politics of Yugoslav 'brotherhood and unity' to advocate greater democratic pluralism and political reforms (Skrbiš 2007b: 226). In this period, one individual who played a catalytic role was the historian Franjo Tuđman. Tuđman always considered the Croatian diaspora to possess an important voice, both in Croatia and abroad and, therefore, obtaining a 'yes' from the diaspora members meant facilitating his ambitions at home.[1]

As followed, every new visit that Tuđman paid to the Croatian diaspora was a step further than a simple confirmation of the previously established co-dependent relationship. And indeed, '[t]he thoughts he had been pondering had coalesced into an ideological vision' (Hockenos 2003: 50). For émigrés he was a hero and for him the émigrés represented primary backing for his policies. They agreed that the Serbs represented the 'arch enemies: We were oppressed by

Serbs, by the Yugoslav army, by Yugoslav diplomacy, Yugoslav trade, Yugoslav commerce, the Yugoslav banking system, Yugoslav organizations, Yugoslav domination' (ibid.: 56). In addition, the idea of creating Greater Serbia was a matter of concern both among the émigrés and the homeland political elite. Given the circumstances, the ethnic Croats insisted that an independent Croatia was a better option than the preservation of a united Yugoslavia. However, limiting this understanding to diaspora groups did not guarantee the desired outcome; instead, the members of diaspora had to take further steps aimed at reaching and convincing policy-makers that international recognition of their independence was the only solution to the Yugoslav crisis. By opting for this scenario, the émigrés took an indirect part in the war, whereas Europeans were trying to understand what was exactly happening and what policy to adopt.

Diaspora strategies within the European Community

The European Community was largely uninformed about the outbreak and rapid development of the Yugoslav state crisis. Often, this ignorance was justified by the assumptions that the Europeans were busy with their own integrationist project which marginalized the situation in Yugoslavia. Even after the outbreak of the hostilities, as noted by Warren Zimmermann, 'the Europeans couldn't believe that Yugoslavia was in serious trouble.... [T]heir approach to Yugoslavia was without any of the urgency with which they acted fourteen months later, when the breakup they said couldn't happen was upon them' (Zimmermann 1996: 65). In fact, it was the US Mission to the European Communities that kept the Europeans informed of what was happening in Europe. The official documents show that daily correspondence between the USA and the European Union (EU) was often unilateral. This kind of communication – which frequently resembled more a brief course on current affairs rather than an exchange of positions about the Yugoslav crisis – created a state of uncertainty. Such a situation, characterized by a lack of engagement and thoughtful policy, provided an opportunity for diaspora groups to react. For them, approaching the Washington or Brussels administration, or both, signified approaching and influencing policy-makers.

The scholarship on the European response to the Yugoslav crisis offers different points of view on diaspora activism. In her study, Beverly Crawford examines the situation in Germany, and while acknowledging the existence of various lobbying attempts, she concludes that diaspora members did not exert any significant societal pressure that could have led to the policy of recognition (Crawford 1996: 502–3). Such a conclusion is dictated by the author's decision to limit her account to the Croatian diaspora only and more specifically to the Croatians residing in Germany, a country where great confusion about the future of Croatia dominated public debates (primarily due to numerous mixed marriages between Croats and Serbs). By contrast, other authors agree upon the diaspora's capacity to raise awareness of the situation in the homeland, push for a rapid resolution and, indeed, affect policies. For example, Sean Carter analyses the

Croatian diaspora and identifies three main areas of diaspora activism which made a difference: fundraising, political protest and public relations campaigning (Carter 2005: 57). In fact, while many accounts of the Croatian diaspora suggest that the Croats were very active abroad, although less successful across the European Community than in North America, Slovenian activism abroad has rarely benefited from any deep analysis.

Slovenian groups gathered around the Slovenian World Congress, a global society including organizations and conferences from all over the world.[2] Even though the Congress was active in different fields,[3] during the Yugoslav crisis, the common objective of securing international recognition of independence was given priority. One study assessed the link between the diaspora and the Congress: 'The Slovenian World Congress was founded and Slovenian politicians repeatedly claimed that the minorities in neighboring countries and the Slovene Diaspora should play a role as "ambassadors" of the new Slovenian state' (Busch 2003: 137). This ambassadorial tactic was not something completely new for the diaspora Slovenes: it had already been embraced in the course of the previous decade. The crisis in the homeland served as an additional motive to adjust their strategy and this time to go a step further than in the recent past.

In 2002, the Slovenian authorities decided to publish archive materials covering the period of the Yugoslav crisis. Unlike other republics that still prefer to have most of their official materials remain well-protected and almost unavailable for research, Slovenia has offered up various collections of documents that facilitate our understanding and encourage further interest in the field. For example, the fourth volume of the *Viri o demokratizaciji in osamosvojitvi Slovenije* [*Sources about Democratization and Attainment of Independence of Slovenia*] offers various transcripts which clearly reveal how the Slovenian leadership communicated with the diaspora and how diaspora groups engaged with the crisis. Here, among the wealth of material, I draw attention to one letter in particular, drafted by Milan Kučan, the Slovenian leader. While addressing his 'fellow cousins' in the diaspora, Kučan, the first democratically elected president of the Republic of Slovenia, expressed his gratitude for their commitment in the struggle for independence:

> Our main objective is to achieve independence of the Slovenian state, freely decide upon our future and we ask for nothing more .... Recently, I have received numerous friendly and encouraging messages. It is not easy to observe what Slovenia is going through at the moment and succeed in our common objective. It is worth mentioning that our independence will be rather a weak one, if we do not reach economic stability. Therefore, I ask the diaspora to contribute with its knowledge and assistance. I am aware of the fact that most of you would like to see Slovenia not only as an independent state, but also as a democratic and stable country. I am fully aware that in order to reach our objectives, you will continue to support us, like you did before.[4]

Kučan's letter, written soon after the Ten-Day War in Slovenia in July 1991, found instant approval among diaspora members. The nature of the letter was both informative, as it described the situation in the homeland, and indicative, as it underlined the policy that the Slovenian leadership considered appropriate to pursue. Kučan pointed out that the Slovenian objectives should be focused on the present and future status of the country, whereas diaspora groups should take an active part in the process.[5] For many diaspora members, the letter represented a necessary guideline, as many of them had almost no idea of the situation in the homeland or the Yugoslav Federation as a whole. However, as a consequence, the diaspora began to coalesce even more.

In Europe, translated versions of Kučan's letter, according to one Slovenian official, reached EC officials through the Slovenian diaspora: 'For Slovenian people, the lack of a clear policy on behalf of the European Union was an opportunity to try and approach decision-makers whose voice was of particular weight. The diaspora did not hesitate; it just continued fighting for something that seemed to be the Slovenian objective since the Second World War'.[6] Indeed, this comment very much complements the idea that the Slovene Immigrants' Society expressed about diaspora activism; the society demanded that diaspora groups inform host governments 'about the critical situation' in their homeland and urge them 'to help preserve peace in the world'.[7]

In order to illustrate the growing Slovenian drive for independence, I identify various aspects that became even more apparent after the Ten-Day War. First, the Slovenian leadership needed to convince the Europeans that its ambition was justified. As noted in a letter addressed to the international cultural and scientific audience:

> All the Slovenian attempts to establish in all of the Yugoslav territory, criteria of European democracy and culture have been without success. The unity of Yugoslavia, so much spoken about these days by the diplomatic representatives of your countries, in our historical experience and particularly in the light of recent events, seems possible only as a dictatorship'.[8]

By doing so, the Slovenes aimed to convince the Europeans that there was something wrong with the principles the Yugoslav Federation was based upon and that they were the ones trying to spread Western European values throughout the country. However, in their view, only after it had proved that Yugoslav unity was not achievable in conformity with European standards, they decided to go for a different strategy. This standpoint complemented what the Slovenian diaspora had already adopted as its official policy before the actual crisis: indeed, the émigrés did not focus on unity at all, but rather the prospect of Slovenian self-determination.

Secondly, by stressing that the Yugoslav People's Army which attacked Slovenia 'understood the viewpoint of the European and the United States government as an excuse as well as encouragement to try to re-establish a centralist dictatorship and to violently nullify the will of the Slovenian nation for self-determination expressed through free elections', the Slovenian leadership tried to show that the Yugoslavs were undermining Europeans and Americans.[9] This argument, stronger than the previous one, blamed the West for their lack of policy and at the same time forced the West to view itself as naïve. In this respect, both Slovenian and Croatian diaspora groups found themselves in an excellent position: criticizing the West for its non-response allowed them to achieve support in the West for their own already established policy. Although Slovenia and Croatia continuously criticized the Community for being late to deliver a clear policy over the situation in Yugoslavia, Europe's late response was nevertheless beneficial for both republics.

Thirdly, the Slovenian authorities insisted that they had satisfied the essential criteria to become an independent state. Accordingly, the republic was 'ready, through negotiations and assumptions of state functions, to commit itself to reliable and stable partnerships with all other nations and authorities of the international community'.[10] This part of the Slovenian strategy focused on delivering a clear message about Slovenia's capacity to fulfil conditions for independence. In addition, it was stipulated that Slovenia was going to work towards a greater stability 'in a complex region of Europe', whereas its independence was not going to be 'at the expense of the region, but as a positive step forward for the whole region and its constituent partners'.[11] The point here was clear as it argued that independence was not to affect the region negatively, but rather contribute to further progress. Controversial as it was, this strategy did not prevent the spread of violence. As Zimmermann, while examining Slovenian behaviour within the Yugoslav Federation and before the crisis, noted: 'In their drive to separate from Yugoslavia [the Slovenes] simply ignored the twenty-two million Yugoslavs who were not Slovenes. They bear considerable responsibility for the bloodbath that followed their secession' (Zimmermann 1996: 71).

Finally, both the Slovenian leadership and diaspora members relied on emotional appeals in their communication with the Europeans. For example, they asked the Community's governments 'in the spirit of humanism and love of freedom, to help [them] in making the truth known – that today in Slovenia, we are defending the foundations of European culture, human rights and self-determination of nations'.[12] The sufferings that post-Second World War Europe had faced served to inspire the Slovenian leadership to advocate the avoidance of similar disasters in the future. The memories of divisions caused by the cold war were still fresh and most of those directly involved in its dismantlement were facing consequences. The emotional dimension was further stressed by statements such as: 'By supporting the right of Slovenia to live a free, independent, cultural and democratic life in its own state, you will support the spirit [of humanism and love], which is our common home'.[13]

Diasporas and individual European states

The Austrian government demonstrated significant sympathy for their Slovenian diaspora. Although Austria was not an official member state of the European

Community at the time of the Yugoslav state crisis, its close links with the Community and its member states – due to the already advanced negotiations of Austrian EC membership – meant that the Austrian standpoint on the situation in neighbouring Yugoslavia was given significant attention by the Brussels administration. For example, a meeting dated 9 July 1991, when Franz Vranitzky, the Austrian socialist chancellor, gathered leaders of Western social-democratic parties in Vienna served primarily 'to exchange views on the Yugoslav crisis' (Klemenčič 2009: 158).

With regard to the Slovenes in Austria, one study showed that although they faced serious mistreatment during and after the Second World War, when the Slovenian population in the country was decimated from 80,000 to 16,000, the Yugoslav crisis showed that the Austrians had since completely reversed their approach towards the Slovenian diaspora members (Cox 2005: 169). In addition, towards the end of the 1980s, both Slovenia and Croatia worked towards securing a stronger link with Austria and Italy through their joint involvement in a regional organization named Alps-Adriatic Working Community, with special focus on economic and political cooperation. In her analysis, Susan Woodward acknowledges the relevance of this venture and stresses that the policies of the organization additionally encouraged the Slovenian and Croatian drive for independence (Woodward 1995: 159). Such a climate provided a favourable environment for diaspora activism. With regard to this, the Carinthian Slovenes played an instrumental role in spreading information about both the Slovenian and the Croatian situation at home and, accordingly, their ambitions abroad (Bahovec 2003).

Due to the geographical, historical and trade proximity, the Slovenian leadership considered Carinthia, a state of Austria, to be the most relevant starting point. One of my respondents acknowledged that communication between them intensified on two occasions: first, in the late 1980s when the Slovenian economy became affected by the economic problems that the Yugoslav Federation as a whole was facing, and secondly, during the Slovenian struggle for independence. Although, according to the source, 'Slovenian complaints in the very beginning focused on blaming the Yugoslav leadership for domestic mismanagement, rather than asking for support in their fight for independence', it soon became obvious that 'all communication was about Slovenian independence and nothing more'.[14] The Carinthian Slovenes responded by providing Slovenian democratic forces with financial aid[15] and by organizing a protest on 27 June 1991 in front of the Yugoslav consulate in Klagenfurt, the capital of Carinthia.[16] This protest, which took place immediately after the outbreak of the Ten-Day War, presented self-determination as the only solution for both Slovenia and Croatia.

The initial Austrian position expressed by Vranitzky, who visited Belgrade before the 1990 elections in Slovenia and openly supported Yugoslav unity, provoked some concerns among the Slovenes. As a reaction, Dimitrij Rupel, first Foreign Minister of Slovenia, immediately criticized the Austrians' perceived intention of shaping and dominating Slovenia.[17] Indeed, the Austrian establishment faced a clash among different political parties in Austria over the

Slovenian question. In his attempt to defend the Slovenes and undermine Vranitzky's words, Andreas Khol, a member of the Austrian People's Party, stipulated that 'the statement of the socialist chancellor was a stab in the back for Slovenia' and openly supported democratic forces which wanted to reconsider Slovenia's position within Yugoslavia.[18] This approach, which dominated political debate in Austria, was further criticized by Alois Mock, Minister of Foreign Affairs of Austria and another representative of the Christian Democrats, who explicitly supported the principles of self-determination and democratization. As he put it: 'Slovenia's future is in the hands of the Slovenes, our good neighbors.... If the outcome implies attainment of independence, Austria will react in accordance with newly displayed circumstances'.[19] For the Slovenes, this statement meant full support from Austrian diplomacy and, accordingly, Minister Mock's figure was appreciated as one of the most prominent Slovenian spokespersons abroad. Having his approval meant securing the Slovenian voice within the European Community in 1991. Abroad, Mock talked about his country's appreciation for the Slovenian diaspora, Slovenes and Croats, in general, and independence as the only solution for them. In his analysis, Klaus Zeitler summarized his intentions:

> Alois Mock was the main advocate of recognition of both Slovenia and Croatia. He tried to convince the international community to intervene militarily.... He tried to convince the Austrian government to support Slovenia logistically and in other ways. Austria even gave Slovenia loans in order to continue its import and export activities in June and July 1991 (Zeitler 2000: 109).

Whereas most of my respondents wished to remain anonymous and avoid naming individuals they viewed as being in charge of various aspects of the Slovenian diaspora policy, some directly involved actors have offered insightful accounts and talked about individual contributions that supported the Slovenes. For example, Janez Stergar, president of the Carinthian Slovenes' Club in Ljubljana, pointed out that Karel Smolle, representing the Austrian Green Alternative Party, Reginald Vospernik, spokesperson for the National Council of Carinthian Slovenes, and Janko Zerzer, head of the Christian Cultural Association, together worked on the Slovenian project by fostering Slovenian diplomatic efforts vis-à-vis European leaders.[20] For all of them, the target consisted in approaching the officials in Brussels. While criticizing the European Community for being in 'a profound dream', Smolle insisted on a rapid resolution. His standpoint was of crucial relevance for both Austrian and European Community authorities. The Austrian leadership, while seeing Smolle's political involvement in line with his ethnic background, thus in favour of the Slovenian diaspora and independence of Slovenia, preferred avoiding any sort of confrontation with diaspora representatives that could have frustrated them and led to sabotage of Austria's membership of the Community (Klemenčič 2009: 158). In Brussels, in order to make the Europeans comprehend the actual gravity of the situation,

Smolle repeatedly compared the situation in Slovenia to the one in Kosovo during the 1980s when the Serbian authorities oppressed Kosovo Albanians. At the time, the Community strongly objected to the Serbian policy in Kosovo, which now seemed to be repeating itself in Slovenia. Thus, having numerous unresolved problems that extended over the whole decade, Smolle described the Yugoslav leadership as in need of expressing its frustration somehow and somewhere; accordingly, it decided to attack Slovenia.[21]

As noted in an interview with one Austrian diplomat serving in Paris at the time, daily statements drafted by diaspora organizations and press significantly shaped the overall perception of the developments of the crisis and ensured that decision-makers within the Community remained updated. The relevance of the diaspora voice is further understood if Viktor Meier's criticism of the Western diplomatic corps and their misunderstanding of the reality of Yugoslavia is taken into consideration: 'I must admit that the views which I heard from the circle of Western diplomats at this time made an almost traumatic impression' (Meier 1999: 21). Thus, within an absence of clear policy on behalf of the official representatives, diaspora representatives were assigned an influential role.

In the interplay between diasporas, homeland and hostland leaderships and decision-makers, it was the National Council of Carinthian Slovenes which repeatedly called upon the Austrian government to recognize the independence of the Slovenian and Croatian republics. On various occasions, the Council insisted on 'urgent recognition' of the republics as a necessary step forward after they had declared their desire to leave Yugoslavia.[22] In his statement, Marijan Pipp, General Secretary of the Council, criticized the fact that the Austrian leadership paid too much attention to the information coming from Belgrade, thus ignoring the real situation across the Yugoslav Federation. Based on this, the Council invited the Austrian leadership 'to take a clear standpoint in that respect and strengthen its economic relations with Slovenia and Croatia and that way help less developed regions of Carinthia and Styria'.[23]

The Slovenes insisted on their regional engagement within the remaining parts of the Yugoslav Federation as well as within Austria's less-developed regions. Finally, the Council approached Minister Mock:

> The National Council of Carinthian Slovenes ... is calling the Austrian government to exploit all diplomatic means in order to prevent the bloodshed in Slovenia and Croatia ..., to consult with the United States and member states of the European Community and that way make them change their policy towards Slovenia and Croatia. Furthermore, the US and EC states should be aware of their responsibility for the situation that has developed so far – a situation they helped develop by their ignoring policy towards Slovenia and Croatia.[24]

The majority of the documents that are available for research were drafted in the summer of 1991 and they all focus on the suffering caused by the Ten-Day War and further advocacy of international recognition of Slovenian and

Croatian independence. Again here, the trend to speak in both republics' name continued: both the Slovenes in Austria, and the Austrian representatives abroad, perceived an independent Croatia as a buffer zone between Slovenia and the problematic Balkans. In their resolution, major Slovenian organizations in Austria insisted on recognition:

> We expect the governments that have already been informed about the situation in Slovenia and Croatia to recognize our independence. In addition, we expect recognition from stubborn governments who insist on Yugoslav unity at any cost, thus against the will of people who participated in its establishment.... Here, we think of the US and Member States of the European Community.[25]

Having argued that the USA and, more importantly for us here, the European Community, lacked a clear and coherent standpoint on the question of Yugoslavia's unity, the Slovenian diaspora triggered doubt among European leaders. It continuously criticized their lack of attentiveness. In Brussels, EC officials knew that something was going on in Yugoslavia, but not many of them knew exactly what. Slovenian insistence on a clear EC policy meant that the officials had to sit down and study the situation in the republic and reconsider contrasting standpoints some of the member states had already developed. By doing so, the Slovenian leadership used diaspora organizations to push for additional attention at a European level.

As a result of the outbreak and outcome of the Ten-Day War, diaspora groups openly challenged both the understanding of democracy in Europe and the aim of European integration if the Yugoslav People's Army was, in their view, permitted to intervene and thus become an army of occupation.[26] Accordingly, Andrej Wakounig, president of the Unity List, a Carinthia-based party which replaced the Club of Slovenian Local Councillors, while stipulating that the attack of the Yugoslav army 'humiliated the Slovenian nation and its young democracy', accused both the USA and the EC of 'uncivilized and malicious reservedness', implying that they had indirectly stimulated the attack through their inadequate approach towards the Yugoslav crisis.[27] The Unity List maintained that Americans and Europeans were responsible for the conflict in the homeland. This pressure for resolution was further justified due to the persistent support the Unity List had shown for Austrian membership of the European Community, which at this particular point appeared to be compromised. The List questioned the point of joining the Community, an organization which does not care about its citizens: 'We do not want to join a group of states which, for their own selfish reasons, allow aggression as a response to Slovenian democratic decisions'.[28]

Other than Austria, another European state where Slovenian diaspora groups tried to rely on a cosmopolitan approach to policy-making and thus speed up the process of recognition was Italy. With regard to the Italians, the resolution drafted by Slovenian organizations in Austria, despite being conceived in Carinthia, was simultaneously published in Trieste. In fact, there is nothing strange about this as the Slovenes and their diaspora representatives actually called on both Austria and Italy to recognize the independence of both Slovenia and Croatia. The Slovenian diaspora in Trieste published a newspaper article entitled 'Occupied Slovenia' in which Marij Čuk reported the attack of the Yugoslav army.[29] Consequently, the members of the Local Council referred to it as an 'unacceptable attack', and called on the international community to step in and help both Slovenian and Croatian republics in their attainment of independence.

Aware of the risk that an immediate Italian decision to recognize the two republics might encourage the further spread of conflict, the members of the Council called on the European Community to prevent military attacks and only then, once military conflict seemed impossible, they also asked the Community to support the independence of the two republics.[30] In Brussels this strategy was perceived as a direct provocation for two reasons. First, expecting a prompt response was impossible due to internal disagreements regarding the Yugoslav crisis. As already noted elsewhere, policies varied significantly among the member states and, in fact, most EC ambassadors expressed reluctance towards the idea of Slovenian and Croatian independence as a necessary precondition for resolution of the crisis. Secondly, the European Community had no power to prevent military attacks; even if it had opted to rely on its member states to step in, it would have eroded the situation both in Yugoslavia and Brussels.

However, for the Slovenian diaspora community in Italy it was important to target the Italian political elite that held the presidency of the European Community at the time (Minahan 1998: 247). This way the Slovenes in Italy achieved a direct connection with Brussels, which facilitated further communication. According to one representative of the Community, Italians tried to coordinate as many sources as possible in order to understand who was trying to obtain what and propose a solution, rather than thinking about a definitive solution for the Yugoslav Federation. In this situation, diaspora groups were well aware of their own contribution. For example, most Slovenes in Italy were part of one of two main groups: the *Krizni štab Slovencev v Italiji* and the *Svet slovenskih organizacij*, both acting as intermediaries between the homeland, diaspora and decision-makers.

Although information about the activism of the first group has remained rather limited, Marko Kosin, head of the Slovenian Bureau in Rome, pointed out that its role was twofold: first, to spread the Slovenian voice on the route from Ljubljana to Brussels via Rome; and secondly, to collect financial aid for what was already known to be 'a new state' (Kosin 1998: 59). Indeed, communication with Rome and Brussels was smooth. In their letter addressed simultaneously to Giulio Andreotti, Prime Minister of Italy, Gianni de Michelis, Foreign Minister of Italy, and Jacques Poos, European Community representative, a group of major

Slovenian organizations,[31] while expressing deep concern over the unexpected hostilities in the homeland, demanded that both Italian and EC officials take an active role in resolving the Yugoslav crisis, precisely to recognize Slovenian and Croatian independence and *conditio sine qua non* for further talks and crisis resolution within the Federation.[32] Here, in addition, the Slovenes insisted that the presence of the Yugoslav army in Slovenia represented an indirect threat for Italy.[33] This suggestion, which was confirmed in Trieste during demonstrations on 30 June 1991, aimed to bring about the following: an immediate end to fighting; new talks based on respect for Slovenian and Croatian independence; understanding and support from the international community; Italian and European recognition; and, finally, full respect of the Osimo Agreements.[34]

The *Svet slovenskih organizacij* acted as a direct external supporter of the Slovenian government. In her letter to the homeland, Maria Ferletič, president of the organization, expressed both sorrow for the violence in the streets of Ljubljana and happiness that the same violence ended quickly and allowed the Slovenes to continue working towards their European goal.[35] In other words, the homeland was obtaining moral support for adhering to its initial policy, which was further complemented by financial assistance. Here, the diaspora thought that the homeland had more chance in the fight for independence if provided with adequate financial contributions – an understanding based on the assumption that financial aid was necessary to secure additional arms if necessary, to cover unexpected official trips and to sponsor various publishing materials about the reality of the Slovenian situation.

Although predominantly active in North America, the activism of the Slovenian World Congress followed the demonstrations in Trieste and offered its full support to the diaspora members within Europe. It proclaimed the week of 8–12 July 1991 to be the week of solidarity during which different Slovenian associations would intensively lobby for their homeland. The Congress advocated that, in different ways, the Slovenes had to try to influence public opinion to recognize the independence of Slovenia. In order to succeed in this, it stipulated: 'We should get the best experts with contacts in different ministries and institutions'.[36]

Indeed, if analysed together, the *Krizni štab Slovencev v Italiji*, the *Svet slovenskih organizacij* and the Slovenian World Congress shared a common objective: an independent homeland. However, as maintained by the Slovenian minorities in Austria, Italy and Hungary, who were desperately hoping for and looking forward to the independence of their homeland as 'the best guarantee to avoid new attacks against Slovenia', continuation of good relations with all neighboring countries was crucial.[37] Regardless of the general strategy adopted by the Slovenes – to fight for both their statehood and that of Croatia – commitment to the surrounding countries always appeared to dominate diaspora discourse, as it was believed that advocacy of strong ties with them would contribute to the European policy that would soon after be formulated.

Conclusion

As demonstrated throughout the chapter, diaspora activism is very much about communication. With regard to the Yugoslav state crisis, successful activism required a well-developed strategy aimed at approaching targeted groups and contributing to policy-making. At EC level, Slovenian diaspora groups played a far more important role than any other republic's diaspora members. While this may seem strange, especially if we think of the size of the Croatian diaspora, the Slovenian vision of the future of the wealthiest Yugoslav republic as an independent state was more established and better conceptualized than that shaped by the Croats. While for the Slovenian diaspora, self-determination of their homeland appeared to be the only acceptable outcome, the Croatian diaspora occasionally appeared divided, as their homeland often included a significant Serbian component. However, this did not prevent the Slovenian diaspora from advocating international recognition of both Slovenia and Croatia.

In Europe, the Slovenes relied on direct support from the Carinthian Slovenes, who had strengthened their political relevance within the hostland while acting as advocates of independence. Therefore, Austria's initial and short advocacy of Yugoslav unity, vehemently criticized by the Slovenes, who believed that immediate support for independence might help avoid further conflict, had nothing to do with diaspora activism. Regardless of the Austrian official standpoint, initial, intermediate or final, the diaspora members worked towards the achievement of their objective. Their work was acknowledged by Janez Dular, the Minister of Slovenian Diaspora, who expressed thanks to the Carinthian Slovenes for 'spreading information among the Austrian public and government'.[38] Once the Austrians said 'yes' to Slovenian and Croatian demands, activism spread outside Austria. On two occasions, the Italian government demonstrated sympathy and its intention to recognize Slovenia and Croatia as independent states: first, in July 1991, during a parliamentary debate;[39] and secondly, in September 1991, when the Slovenian diaspora arranged for 20,000 postcards exclaiming 'Yes to Slovenian and Croatian Independence!' to reach Andreotti's cabinet.[40]

The Austrians and Italians played a significant role as they encouraged other member states of the European Community to take the situation seriously within the two republics, and in Yugoslavia in general. As observed by Vranitzky after the EC Ministerial meeting on 16 December 1991, 'in reality, the EC Foreign Ministers decided to do what we had decided two-and-a-half months ago on principle, namely to recognize Slovenia and Croatia' (Grant 1999: 188). The EC's decision to finally recognize them on 15 January 1992 can be interpreted as going hand in hand with the idea of a new cosmopolitan Europe. This Europe discredits the concept of national solidarity as inadequate to tackle highly sensitive issues of international relevance and thus promotes transnational solidarity (Beck and Grande 2008: 189–91). Aware of the complexity of the Yugoslav drama, the Europeans opted for a policy which in their view represented the only acceptable solution, hoping to boost the profile of their new, post-cold war and, more precisely, post-Maastricht Europe.

Notes

1 For example, Tuđman's 1987 visit to the Croatian diaspora in North America was beneficial for three reasons: first, he established a strong linkage between the pro-independence movement in Croatia and the diaspora; secondy, he secured financial support for his campaign back home; and thirdly, he developed a co-dependent relationship between the diaspora and the homeland.
2 American Slovenian Congress, Canadian Slovenian Congress, German Slovenian Congress, Slovenian organizations in Italy, Austria, Australia, Argentina and Great Britain and Carinthia Conference.
3 Promoting Slovenian identity and language, securing mutual support, fostering cultural linkages, encouraging equality, promoting Slovenian national interests both at home and abroad, integrating all Slovenes, promoting tolerance and respect.
4 'Milan Kučan's letter to the diaspora', 22 July 1991.
5 Similar points were made in correspondence drafted in various ministries. For example, Jelko Kacin, Minister of Information, on headed paper bearing the logo Independent Slovenia 1991, stipulated that his Ministry focused on informing the international audience about the situation at home. That way, diaspora organizations had an opportunity to shape their activism (Jelko Kacin's 'Letter to the émigrés', quoted in *Viri o demokratizaciji in osamosvojitvi Slovenije*, 25).
6 Interview with a Slovenian liberal politician, 2007.
7 Letter from Mirko Jurak, President of the Slovene Immigrants' Society, to all Slovenes entitled 'An appeal for support', Ameriška domovina, 11 July 1991.
8 Letter to the international cultural and scientific audience, 28 June 1991', *Viri o demokratizaciji in osamosvojitvi Slovenije* (IV del: Slovenci v zamejstvu in po svetu), p. 32.
9 Ibid.
10 'Letter to the Heads of State, Prime Ministers and Ministers of the Commonwealth Nations, 17 October 1991', in *Viri o demokratizaciji in osamosvojitvi Slovenije* (IV del: Slovenci v zamejstvu in po svetu), 34.
11 Ibid.
12 Ibid.
13 Ibid.
14 Interview with a Slovenian liberal politician, 2007.
15 The aid consisted of two cheques: one for 870,000 schilling and the other for 470,000 schilling. See 'Čez milijon šilingov pomoči', *Slovenski vestnik*, 25 September 1991.
16 The protest, the main objective of which was to condemn the Ten-Day War in Slovenia, gathered members of the Slovenian organizations in Carinthia, including the National Council of Carinthian Slovenes, Association of Slovenian Organizations, Slovenian Unity List, Christian Cultural Association, Slovene Cultural Association. The participants relied on the principle of self-determination and called on the international community to support Slovenian and Croatian bids for independence.
17 *Delo*, 6 April 1990.
18 '*Delo*, 19 April 1990.
19 *Delo*, 14 March 1991.
20 Stergar, Janez, 'Delovanje Koroških Slovencev v Avstriji za neodvisno Slovenijo', in *Viri o demokratizaciji in osamosvojitvi Slovenije* (IV del: Slovenci v zamejstvu in po svetu), 38.
21 'Pomoč Evrope k emancipaciji: drugačno mišljenje, ki ne bi zagotovilo prihodnosti samo Sloveniji', *Naš tednik*, 26 July 1991.

22 'Narodni svet koroških Slovencev zahteva da Avstrija prizna samostojnost Slovenije: Tudi minister Busek podpira prizadevanje mladih demokracij', *Naš tednik*, 1 March 1991.
23 Ibid.
24 'Letter to Dr Alois Mock on behalf of the National Council of Carinthian Slovenes', 27 June 1991.
25 'Resolucija udeležencev zborovanja za spoštovanje pravic narodov do samoodločbe in protest proti vojaškemu nasilju v Sloveniji i Hrvaškem', *Primorski dnevnik*, 28 June 1991.
26 'Izjava glavnih odborov Zveze slovenskih organizacij na Koroškem in vključenih organizacij ob državni osamosvojitvi Slovenije in vojaškem posegu jugoslovanske armade', *Slovenski vestnik*, 3 July 1991.
27 'Zahteva Andreja Wakouniga, predsednika Enotne liste – EL, za avstrijsko priznanje državne samostojnosti Republike Slovenije', *Naš tednik*, 5 July 1991.
28 Ibid.
29 'Slovenija okupirana', *Primorski dnevnik*, 28 June 1991.
30 In Consiglio comunale, in *Viri o demokratizaciji in osamosvojitvi Slovenije* (IV del: Slovenci v zamejstvu in po svetu), 66.
31 Slovenska kulturno-gospodarska zveza, Svet slovenskih organizacij, Organizacije videmske pokrajine, Slovenska komisija PSI, Slovenska komponenta DSL, Slovenska skupnost, Gibanje za komunistično prenovo.
32 'Ustaviti agresijo', *Primorski dnevnik*, 29 June 1991.
33 'Izjava Slovenske deležne komisije', *Primorski dnevnik*, 29 June 1991.
34 *Primorski dnevnik*, 1 July 1991. The 1975 Osimo Agreements settled the territorial dispute between Italy and Yugoslavia and led to greater bilateral cooperation.
35 'Pismo Sveta slovenskih organizacij predsedniku vlade Republike Slovenije Lojzetu Peterletu', *Primorski dnevnik*, 28 June 1991.
36 'Poziv deželnega odbora Svetovnega slovenskega kongresa za Furlanijo-Julijsko krajino', *Primorski dnevnik*, 9 July 1991.
37 'Skupna izjava treh slovenskih narodnih manjšin', *Primorski dnevnik*, 14 July 1991.
38 'Zahvalno pismo Ministra Dr. Janeza Dularja koroškim Slovencem za dveletno pomoč v osamosvojitvenih prizadevanjih Republike Slovenije', *Naš tednik*, 24 January 1992.
39 'Predstavniki manjšin v poslanski zbornici zahtevali priznanje neodvisnosti Slovenije', *Primorski dnevnik*, 5 July 1991.
40 'Prva pobuda združenja Italia-Slovenia. Dvajset tisoč razglednic Andreottiju za priznanje Slovenije in Hrvaške', *Primorski dnevnik*, 29 September 1991.

4 Diaspora, cosmopolitanism and post-territorial citizenship in contemporary Croatia

Francesco Ragazzi

When it gained independence in 1991, the newly born Croatian state enacted an unprecedented set of policies to reach out to its population abroad. It was based on a radically new vision of the Croatian nation promoted by Franjo Tuđman, leader of the HDZ (Hrvatska Demokratska Zajednica – Croatian Democratic Union) and freshly elected first president[1]. The Homeland Croatia, Tuđman claimed, could only be complete if reunited with the Emigrant Croatia, composed of Croats dispersed around the world. Based on this, the new government, both in the constitution and in the law on citizenship, enshrined a new, transnational conception of the Croatian nation. It provided Croats abroad with the right to vote and be elected in a separate electoral unit in the Croatian Parliament, and established several institutions – including a short-lived ministry – to actively promote and encompass the Croatian diaspora as a full member of the nation.

While Croatia was an isolated case in those years, 20 years later it appears as the precursor of a much broader trend. Not a year passes without a new government joining the trend of what Alan Gamlen (2009) has termed 'diaspora engagement policies,' ranging from the symbolic politics of diaspora inclusion such as the organization of dedicated conferences and conventions or the institutionalization of economic and political networks through consular and consultative bodies, to full-blown ministries and administrations for the diaspora. China, Haiti, India, Ireland, Israel, Mexico and Morocco are only some of the most advanced states engaged in these practices of incorporation of the diaspora symbolically, materially and politically.

These developments do not go without questioning some of the most fundamental assumptions of the Westphalian project of the territorial state, and in many regards of the liberal theory of representative politics that is associated with it. More precisely, transformations such as the tolerance towards dual citizenship, the possibility of obtaining citizenship abroad, or voting and being represented as a constituency abroad challenge the traditional assumptions of a match between identities, borders and social and political orders (Albert *et al.* 2001). They assume and enact the conception that the 'domestic is not necessarily within the borders of the nation-state, but can in fact be a 'domestic abroad' (Varadarajan 2010).

What is the meaning of these transformations, and what have been the main interpretations of it? Interestingly enough, the independence of Croatia – while marking the beginning of a tragic decade of violence in the Balkans – corresponded worldwide with an opposite trend: namely, the collapse of the bipolar confrontation, the emerging perception of 'globalization' as a distinct economic and social phenomenon, and with it, the hopes of a political order that would overcome the traditional frontiers of the nation-state. The end of the cold war has in particular generated a renewed interest in ideas of cosmopolitanism and brought into existence a vibrant interdisciplinary field of study concerned precisely with the possibility of thinking citizenship beyond the territorial state. And so it is not surprising that it is predominantly within this framework that diaspora policies have been interpreted.

The aim of this chapter is, however, to question this optimistic reading of diaspora policies. I contend that if cosmopolitanists are right in finding in these developments the emergence of political spaces beyond the territorial state, it is, however, a fundamental mistake to conflate them with manifestations of progressive or cosmopolitan politics. It is, indeed, important to distinguish the post-territorial nature of diasporic practices of citizenship – what I term post-territorial citizenship – with the misleading, normatively charged assumption that they are necessarily a form of cosmopolitan resistance to an exclusionary or repressive order. In fact, in the Croatian case, the opposite is true: diasporic citizenship has been ethnicized from the onset and has gone hand in hand with exclusionary practices of citizenship towards non-Croats on the territory, organizing what Igor Štiks and I have defined as the 'ethnic engineering of citizenship' (Ragazzi and Štiks 2009).

The first section of the chapter unpacks the notion of post-territorial citizenship against the backdrop of the current literature on cosmopolitanism and cosmopolitan citizenship. I then move to the analysis of the legal and institutional arrangements that form the core of Croatia's post-territorial citizenship.

Cosmopolitan, post-national or post-territorial citizenship?

Why are governmental enactments of diasporic citizenship – i.e. practices of extending civic, social and political citizenship rights to populations abroad – not necessarily markers of cosmopolitan citizenship? At first sight, diasporic citizenship seems to have many affinities with cosmopolitan citizenship. As Andrew Linklater argues, the cosmopolitan conception of citizenship stands against 'traditional perspectives [which] maintain that modern conceptions of citizenship are anchored in the world of the bounded community; [which] contend that it loses its precise meaning when divorced from territoriality, sovereignty and shared nationality' (Linklater 1998: 22). It puts forward a notion of citizenship that takes into account the changes brought about by the flows of people, capital and ideas that crossnational boundaries at an increasing pace and therefore posits a global sense of entitlement and responsibility that is not and should be bound by the territorial borders of the state (Linklater 1998: 23).

In this framework, migrants and diasporas are the primary political subject of cosmopolitan change and hybridization (Appiah 2006; Clifford 1994; Gilroy 1994). In the normative opposition between a closed, exclusionary nation-state and an emancipatory transnational alternative, diasporas are the ontological contenders, the cosmopolitan challengers of the nation-state (Appadurai 1991; Cohen 1996).

And so, rather unsurprisingly, cosmopolitanists have located state-backed diaspora policies in this framework. As Seyla Benhabib puts it:

> today nation-states encourage diasporic politics among their migrants and ex-citizens, seeing in the diaspora not only a source of political support for projects at home, but also a resource of networks, skills and competencies that can be used to enhance a state's own standing in an increasingly global world [...].
>
> (Benhabib 2007: 24)

Transnational migrations, Benhabib explains, foreground the 'pluralization of sites of sovereignty' in that diasporas live across multiple legal systems and jurisdictions, under the protection of cosmopolitan norms enforced by human rights treaties – and in struggle with a system of state sovereignty which privileges national citizenship and restricts dual and multiple citizenship (Benhabib 2007: 24). Hence, diaspora policies support the 'cosmopolitan argument' that

> world citizenship can be a powerful means of coaxing citizens away from the false supposition that the interests of fellow citizens necessarily take priority over duties to the rest of the human race; it is a unique device for eliciting their support for global political institutions and sentiments which weaken the grip of exclusionary separate states.
>
> (Linklater 1998: 22).

Aihwa Ong has already produced an extensive critique of the 'innocent concept of the essential diasporan subject, one that celebrates hybridity, "cultural" border crossing, and the production of difference in the work of Gilroy, Hall, Clifford and Bhabha' (Ong 2000: 13). I will therefore not rehearse it here. The main issue I take with existing explanations is what we could define as the spectrum of 'methodological *post*-nationalism'.[2] The 'post-national' argument has been popularized, among others, by the work of Yasemin Soysal.[3] Soysal argues that migrants and diasporas increasingly escape the statist conception of the nation through subnational, supranational or transnational forms of belonging, anchoring their rights as citizens no longer in the institutions of the state, but through forms of 'universal personhood' embedded in novel human rights norms and treaties which came to force after the Second World War. In doing so, they are 'post-national', and therefore cosmopolitan.

By 'methodological *post*-nationalism' I therefore want to suggest, as the proponents of 'methodological nationalism' do,[4] that much of the literature on citizenship and diaspora politics has been permeated by an untold, unconscious set of normative and methodological assumptions that need to be made explicit in order to be overcome. The first problematic assumption lies in the conflation of the 'national' with the 'territorial'. All that is 'national' – and therefore nationalism – can only be territorial. Anything that goes beyond the territory is therefore, to the opposite, post-*national*. Thus, transnational practices are interpreted as necessarily undermining the logic of nationalism and its exclusionary effects – and therefore a sign of resistance or cosmopolitanism. The second problematic assumption is that state power is exclusively territorial. Processes that go beyond the territoriality of the state are assumed to somewhat escape its power – and therefore solely ruled by individual (ethical) or global (human rights) norms. The conflation of the nation and territory, on the one hand, and state and territory, on the other, paradoxically reproduce, in reverse, the ontological bias of methodological nationalist social science. Mine is therefore not a statist critique of post-nationalism,[5] but a critique of the core assumption of the territorial boundedness of governmental power in explaining the attention of governments to their populations abroad.

What is the alternative? Elsewhere, I have proposed to look at diaspora policies as a shift in the geography of the practices of governmental power (Ragazzi 2009b), by drawing on a large literature in sociology (Hindess 2001; Sassen 2006), political geography (Agnew 1994, 2009; Dalby and Toal 1998) and international relations (Bigo and Walker 2007; Ruggie 1993), which has now shown that the state is neither resisting nor being undermined by processes of transnationalization. It is in fact becoming transnational itself. And while a subfield of diaspora studies now acknowledges the importance of diaspora policies (Adamson and Demetriou 2007; Gamlen 2009; Levitt and de la Dehesa 2003) and emigrant citizenship (Barry 2006; Bauböck 2005; Fitzgerald 2006; Itzighson 2000), only a few locate them in a process of transnationalization state practices (Gray 2006; Kunz 2011; Larner 2007; Ong 2000; Smith 2003a, 2003b; Tintori 2011;Varadarajan 2010).

Building on the work of this latter group of scholars, I suggest to conceptualize state practices of extending citizenship to its citizens beyond the traditional borders of the nation-state, forsaking territorial criteria of belonging – such as birth or residence – as practices of 'post-territorial citizenship.' In my understanding, post-territorial citizenship is therefore not a handy synonym of post-national citizenship – as have most authors who have used the term so far – but as a term which undermines its very assumptions.[6] The characteristic of this new form of post-territorial citizenship is indeed to abandon the territorial referent as the main criteria for inclusion and exclusion from citizenship, focusing instead on ethno-cultural markers of identity, irrespective of the place of residence. Post-territorial citizenship is hence closely linked to, and partly overlaps with, what Riva Kastoryano and Nina Glick Schiller, among others, have defined as transnational forms of nationalism (Glick Schiller and Fouron 1999; Kastoryano 2007).

As a historical phenomenon, post-territorial citizenship is not new.[7] It takes its roots, and often its model, from ethnic definitions of nationality. But while these states did enact early forms of post-territorial citizenship in the late nineteenth century, they met repeatedly with two strong territorializing processes. They first met a strong resistance at the international level until recently. Since the Bancroft treaties of the 1860s, and the European Convention on nationalities up to 1997, international agreements actively discouraged dual nationality, the sources of which were precisely practices of post-territorial citizenship. Secondly, for the most part of this period, even when the nation was conceived in ethnic terms, such as in Italy, Germany, or Israel, the political goal up until the past few years was always to 'territorialize' it or 'populate the land' (Joppke 2003; Kimmerling 1983). The nationalities of the Soviet Union, even if conceived in ethnic terms, also had to be territorialized in order to be recognized as such.[8] Throughout the twentieth century, whether civic or ethnic, nations were to be *territorialized*. The novelty is that they are not anymore.

Thinking of these practices as post-territorial allows, therefore, two conceptual moves. First, it allows a move from a topology of levels (the subnational, the national, the global) to a topology of transnational, overlapping social spaces (Faist 2004; Pries 2001). Practices that cross territorial boundaries create a transnational site of contestation – and therefore politics – and there is certainly no ontological necessity for these practices to be invested in a particular normative position. Secondly, this space is not a separate, free-floating space of ideas and cosmopolitan norms, but (to the contrary) a space saturated with a multitude of relations of power, legislative systems and bureaucratic routines in which several actors – governments, associations and individuals – are in permanent competition for a material and symbolic definition of the included and the excluded. Analysing enactments of citizenship as post-territorial therefore allows an exploration of the way in which these enactments of citizenship open or close down these political spaces. Therefore, I now move to analyse Croatia's post-territorial citizenship policies.

Croatia: enacting post-territorial citizenship

Contrary to much folk wisdom about Balkan nationalism, the major change brought about by Tuđman and the HDZ in post-Yugoslav Croatia was not the promotion of the 'Herderian' (*jus sanguinis*) conception of citizenship, to the expense of a previously 'Renanian' (*jus soli*) conception. Citizenship in Yugoslavia, as in most countries of Eastern Europe, was almost always based on an ethnic conception of citizenship. The main substantial shift with the independence of Croatia occurred at the level of the geographical imagination – i.e. from a *territorial* to a *post-territorial* conception of the nation. In the following sections, I outline the features of Croatian's post-territorial citizenship. Against interpretations of diasporic citizenship as cosmopolitan, I argue that, in the Croatian case, diasporic citizenship serves as an encoding of exclusionary ethnic politics. Post-territorial citizenship, here, works as a way to re-frame the borders of the nation, privileging ethno-religious criteria at the expense of the territorial ones in vigour during the Yugoslav period.

Citizenship law

As was common in Communist states,[9] Yugoslavia's conception of citizenship was closely knit with the ethnic understanding of the nation. The citizenship of the Federal People's Republic of Yugoslavia (FPRY) – to become the Socialist Federation of Republics of Yugoslavia (SFRY) on 5 July 1956[10] – was first defined by belonging to a particular nation (*narod*), which itself belonged to the Federation. Nations themselves were *territorialized* as constitutive peoples of one of the six republics inside the Federation. Every citizen officially declared a (voluntary) 'ethnic' affiliation,[11] every republic was the republic of one or more particular constitutive nation and every citizen had a republican citizenship. Within the framework of the Federation, the distinctions were irrelevant to everyday life (Štiks 2006).

With the Croatian Declaration of Independence of 25 June 1991, a new citizenship law was approved,[12] changing the link between the 'nation' and 'territory'. The law was conceived on the basis of two principles: legal continuity with citizenship of the Socialist Republic of Croatia and Croatian ethnicity (Omejec 1998: 99). By law, all possessors of the former Croatian republican citizenship became citizens of the new state, making all other residents *aliens* overnight, regardless of the duration of their residency in Croatia. Their naturalization was then regulated by article 8 of the Law on Croatian Citizenship, which requires five years of registered residence in Croatia, provided that the following conditions were met: no foreign citizenship or proof of release from a previous citizenship, if admitted to Croatian citizenship; *proficiency* in the Croatian language and *Latin* script; *conduct* that reflects an attachment to the customs and legal system of the Republic of Croatia; and, finally, acceptance of *the Croatian culture*. These supplementary criteria made the territorial element in the attribution of citizenship irrelevant (Štiks 2010).

Croatian ethnic origin was usually determined through any official document released by SFRY or republican authorities in which one declared oneself to be ethnically Croat, but sometimes peculiar documents such as Catholic Church certificates were also accepted by the state authorities (among Southern Slavs, being a Roman Catholic is considered the strongest proof of someone's 'Croatianness', if born south of Slovenia). The law facilitated the naturalization of emigrants and their descendants who accepted the Croatian legal system, customs and culture, even if they did not match the criteria defined in article 8 (article 12). Moreover, it allowed ethnic Croats without previous or current residence in Croatia to obtain Croatian nationality by declaration through article 16.

This affected not only the historical 'diaspora' in the USA, Canada and Argentina but also Croats from neighbouring Bosnia and Herzegovina. Despite being one of the 'constitutive peoples' of the Republic of Bosnia and Herzegovina, they were considered potential Croatian citizens in 'diaspora' and included in the legislative provisions. Indeed, article 16 facilitated the naturalization of ethnic Croats living in the 'near abroad' (former Yugoslav republics), especially for those in Bosnia-Herzegovina, while article 11 facilitated the naturalization of the Croatian ethnic emigrants and their descendants, even if they did not satisfy the conditions stated in article 8 regarding proficiency in the Croatian language.

While the new law included previously ostracized Croats abroad, it suddenly excluded from citizenship those former Croatian citizens who had lived all their lives in Croatia: mostly ethnic Serbs. The new citizenship law put those with less than five years of registered residence and those who were unable to prove that they had been released from foreign citizenship (i.e. previous republican citizenship) in a particularly difficult position. In a context in which Croatia was at war with the Yugoslav Federation it was virtually impossible to satisfy this condition. Only aliens born in the territory, spouses, emigrants and those whose citizenship is of interest to Croatia did not have to prove release from their previous citizenship under the naturalization procedure. Moreover, all applicants for naturalization had to prove that they accept 'the Croatian legal system' and 'customs' and 'culture'.

The ethnic, post-territorial features of the 1991 Citizenship Law were confirmed in the transitional provisions, determining the initial citizenry of Croatia and including a special mode of acquiring citizenship for ethnic Croats who were registered but did not possess Croatian republic-level citizenship. They could acquire Croatian citizenship by issuing a written statement to the police, that they considered themselves Croatian citizens. Once the police had confirmed that the applicant fulfilled the above requirements, the applicant was entered into the citizenship registry (see art. 30, para. 2).

Since legal continuity with previous citizenship in the Socialist Republic of Croatia was the determining factor for the establishment of the initial citizenry of the newly independent state, the Republican Registrar's Office was supposed to issue certificates on Croatian citizenship. However, problems arose if the person was registered, but his or her republican citizenship was not Croatian (if, for instance, the father's republican citizenship was sometimes used to determine the republican citizenship of the child), or if no republican citizenship was officially recorded. The former cases were considered aliens and had to apply for naturalization, whereas the latter were sent to police agencies to have their citizenship determined or were allowed to register as Croatian citizens – according to art. 30, para. 2 – if they were able to prove Croat ethnic origins (UNHCR 1997). If they were not able to provide the necessary proof or were simply of a different ethnicity, they remained aliens under the law.

For Serbs who had remained in Croatia, the exclusion was not only symbolic but also had a direct impact on everyday life. Individuals could lose their jobs under the Employment Law for Foreigners, which regulated work for the new non-citizens, requiring them to obtain a work permit. Work permits could be issued to foreigners only if they had specific qualifications that were necessary in the country and could not be filled by Croatian citizens. Thus, ethnic discrimination in the workplace could be justified on a massive scale through the new citizenship law (Human Rights Watch 1995: 15).

The citizenship status of Croatia's Serb minority in the Krajina region was even more problematic.[13] After the Croatian Serb militia, with help of the Yugoslav federal army, took control of almost one-third of Croatia's territory, mostly in the Krajina region (but also in Central and Eastern Slavonia) in 1991, the citizenship status of the ethnic Serb population living in these regions remained unresolved for almost a decade. Croatian Serb refugees who fled or were forced to leave Krajina during and after the Croatian military takeover in 1995 found themselves in Serbia or Bosnia-Herzegovina (in the Serb entity) and were in a particularly difficult situation. Legally, they were all Croatian citizens, but they did not possess a certificate of Croatian citizenship (*domovnica*) and so could not claim all of the rights due a Croatian citizen.

Citizenship functioned as one of the main sites in which the post-territorial exclusionary logic of nationalism was expressed, excluding the unwanted present to the profit of the wanted distant. This logic was mirrored in the modalities of political representation.

Political representation

The Constitution of 1990 guaranteed the right to vote to all Croatian citizens regardless of residential status (Croatia 1998 [1991]). Although the HDZ had advocated a diaspora ballot since 1990, it was only on 18 September 1995 that the Croatian parliament was presented with a new election law that included, for the first time, the right to vote for Croats residing outside of the borders of Croatia.[14] The law set up a fixed number of representatives for the parliament, elected on a separate electoral unit and on a separate electoral list. By the introduction of this new legislation, the electoral corpus of Croatia was increased by about 10 per cent, and the 'diaspora' was thus awarded 12 seats out of 151 in the Sabor – the Croatian parliament (Bajruši 2007; Salay *et al.* 1996: iii).

The anticipated 1995 elections were to be organized in advance to capitalize on the victory of the Flash and Storm operations, which allowed for the forced reintegration of the separatist provinces of Krajina and Eastern Slavonia. Observers commented that the granting of voting rights to the 'HDZ-sympathetic' diaspora was a move to further consolidate the party's power in parliament. The move did benefit Tuđman's party: 398,839 voters were registered abroad, and although the turnout was rather low (109,389 or 27.4 per cent) all 12 seats went to the HDZ (Kasapović 1996: 270). Yet the system was criticized on the basis that the mandate on the 'diaspora' list represented fewer people than the average seat (Kasapović 1999; Mecanović 1999).

In 1999, after heated debates, the electoral system was changed.[15] From a fixed number of seats, the law proposed proportional representation based on voter turnout. This again guaranteed the HDZ six seats out of six in the 2000 elections (Zdenka Babić-Petričević, Milan Kovač, Ljubo Ćesić Rojs, Ante Beljo i Zdravka Bušić). This again had an important political impact. Although the HDZ lost overall, the SDP–HSLS coalition (led by Ivica Račan and Stipe Mesić) obtained 47 per cent of the votes instead of the 50.7 per cent majority they would have won without the diaspora seats. Moreover, the extra seats made the HDZ first in the parliament in terms of single-party representation.

The diaspora constituency continued to make and break candidates. In the presidential campaign of 2005, despite the low turnout (19 per cent), the diaspora vote both prevented the election of Stipe Mesić (HNS) in the first round and allowed the HDZ candidate Jadranka Kosor to dispute the second round instead of Boris Mikšić, who would have qualified instead ('Izbori 2005 – Arhiva'). The political importance of the vote was even higher in 2007, when it provided the HDZ with the necessary extra seats to form a government ('Izbori 2007 – Rezultati').

It therefore appears clearly that Croatia's 'diasporic' citizenship policy is deeply embedded in a process of inclusion of ethnic Croats. Yet, the other side of the coin of this post-territorial citizenship policy is that, far from the cosmopolitan, post-national ideals that some authors would advocate, it forcibly excludes from citizenship those non-Croats living in the territory. And the close entanglement of these two features was directly addressed in the debates linked to the question of the right to vote and to be represented in parliament. As one member of the Sabor, elected on the diaspora ticket, explained:

> I am mostly saddened by the discussions that examine the number [i.e. reduction] of representatives from the diaspora. We have been forced out of the homeland because of the circumstances, not only me, but thousands of Croats. We helped to create the free Croatia in many ways, and today we have only twelve members of parliament. [...] On the other hand, no one mentions the members of parliament representing the national minorities, the majority of which, if they haven't taken part in it themselves, have at least advocated for the aggression [i.e. the war] against the Republic of Croatia.[16]

This common argument was in stark contrast with reality. In fact, since the independence of Croatia, the HDZ's policies had reduced the importance of minority representation.

The 1992 law on the rights of national minorities,[17] passed under pressure from the European Community, granted 13 out of 138 seats to minorities in the Croatian Sabor. Since the law allowed only minorities representing more than 8 per cent of the population to be represented, Serbs took all 13 seats. The representation of minorities was calculated on the basis of the 1981 Yugoslav census.[18] Yet, in 1995, the new law erased the reference to the 1981 census, stating that minority representation was postponed until a new census was completed, a technicality to prevent minorities from being represented (Burrai 2008: 13).[19] Moreover, voter participation was tightened and openly discriminated against non-ethnic Croats. While registration was extended to 300,000 out-of-country ethnic Croat voters, 200,000 primarily ethnic Serb voters (5 per cent of the potential electorate) who had fled the country were excluded from the ballot (OSCE 1997: 3). The new law of 1999 had further restricted minority representation, and Serbs only obtained one representative at the Sabor.

In the 2000 parliamentary election, the attribution of minority seats had 'diluted' the representation of the Serb population by allowing other minorities to be represented. One seat was allocated for Serbs, Italians, Hungarians, Czechs and Slovaks together with 'others' (Austrians, Germans, Ukrainians, Ruthenians and Jews) (OSCE 2000: 3). The electoral procedure created further discrimination between the internally displaced 'expellees' and 'displaced persons':

> A large number of polling stations were established for 'expelled' persons who are generally ethnic-Croats, whereas only two polling stations were set up for 'displaced persons', generally ethnic-Serbs. This segregation of internally displaced voters between two categories and the disproportionate number of polling stations provided to each category, in effect, discriminate between voters of Croat and Serb ethnic origin.
>
> (OSCE/ODHIR 2000: 6–7)

However, this progressive reduction of minority representation was reversed with the end of the Tuđman era and the election of Stipe Mesić and the installation of the left-wing Račan-led government. In 2000, a new Constitutional Law on National Minorities was passed, increasing representation of minorities in the Sabor. For the 2003 and 2007 parliamentary elections and the 2005 presidential election, although the HDZ regained power, the provisions remained unchanged.

In sum, the reintegration of Serbs into the body politic was linked to legal issues of citizenship and representation, and the more pernicious reversal of a diffused policy of physically eliminating the minority from Croatia's territory.

Institutions

The central role of the Croatian diaspora in the foundation of Franjo Tuđman's Croatian Democratic Union and during the 1991–1995 conflict in providing weapons, humanitarian aid, lobbying and even soldiers allowed the Tuđman government to provide a particular prominent position of the diaspora in post-Yugoslav institutions (Hockenos 2003; Ragazzi 2009a; Winland 2007).

The first institution created to deal with the diaspora was the short-lived Ministry for Emigration, or Ministry for the Diaspora (*Ministarstvo za Iseljeništvo*), briefly lead by Gojko Šušak in autumn 1991. After years of opposition and exclusion from the regime, Croats abroad had their 'own' Ministry.[20] The second ministry was set up under the Zlatko Mates government and named the Ministry for Return and Immigration (*Ministarstvo Povratka i Useljeništva*). From November 1996 to June 1999 it was led by Marijan Petrović, until in June 1999 under Jure Radić, it became the Ministry for Development, Immigration and Reconstruction (*Ministarstvo razvitka, useljeništva i obnove*), from May 1999 to January 2000. It included a special department for return, led by Marin Sopta. The minister's cabinet was mostly composed of former Croatian emigrants, either active politically[21] or in emigration-related cultural associations and journals.[22]

The main goal of the ministry was to create the conditions for a Croatian return of the diaspora – similar to the Jewish conception of *Aliyah* – to take place. Several measures were taken, such as reducing tax requirements for returnees, a 25 per cent cap on the tax on foreign pensions, a facility for the return of capital, and special agreements for health care or pension. Moreover, informative material was published, such as a guide for returnees. Some initial promises were only partially fulfilled, such as special assistance for administrative matters, an insurance plan to fund return plane tickets for poorer migrants, and the opening of a fund to provide for returnees without means of subsistence. The ministry also planned different forms of scholarships, such as a scholarship with the Ministry of Defence for young English-speaking Croats who wanted to return.[23]

In addition to these measures, the Ministry was to collaborate with other Croatian institutions: with the diplomatic services to set up special counsellors for immigration; collaboration with *Hrvatska Matica Iseljenika* and the Croatian Information Centre to produce statistical information on the number and condition of the returnees and potential returnees; with the Ministry of Interior to facilitate the acquisition of Croatian citizenship.[24]

With the new government, the Emigrant Foundation of Croatia (*Matica Iseljenika Hrvatske, MIH*) was renamed the Croatian Heritage Foundation *(Hrvatska Matica Iseljenika, HMI)*. Matica's objective continued to be the preservation of the 'homeland identity' abroad. The main difference was that Matica shifted from the promotion of the socialist self-management identity of the Yugoslav years to the promotion of a nationalist identity. Throughout the 1990s, by becoming entirely subordinated to the HDZ, *Matica Hrvatska Iseljenika* fell again into the function it had during the years of communism: an uncritical, ignored organ of the established power. With the election of the new majority in 2000, the institution progressively de-politicized and concentrated on underfunded programmes of cultural and educational exchange. It was conceived as an institution to organize Croatian lobbying and to coordinate institutions abroad, but Matica never fulfilled its expectations.

Finally, a last institution that tried to develop organizations abroad as a resource was the Croatian World Congress (*Hrvatski Svjetski Kongres, HSK*), launched in July 1993.[25] Initiated by a group of moderate emigrants, it (again) was rapidly taken over by the HDZ, under the leadership of Franjo Tuđman who appointed the Herzegovinian Franciscan friar Šimun Šito Ćorić at its head. The explicit goal was to set up a structure that could represent all the emigrant associations under one roof and to reinforce their activities abroad. However, in most places, the clumsily obvious desire to control the entire Congress, and deal exclusively with organizations friendly to the HDZ, created more division than unity.

Similarly, institutional arrangements symmetrically excluded non-ethnic Croats in various aspects of everyday life. We outline only three of the most pressing issues in this regard. The first one regards return policies. In 1997, a return policy (primarily for Croatian Serbs who had escaped during the war) was implemented in the section of Slavonia still temporarily occupied by international forces (UNTAES). Besides being enshrined in international law,[26] the principle of return was explicitly agreed upon by Croatia in the Dayton Peace Accords. In theory, all displaced persons (Serbs) were authorized to stay in the houses they occupied before they could return to their original houses (usually occupied by Croats, often themselves refugees from Bosnia-Herzegovina). But after the UN withdrawal, thousands of Croats returned and claimed their possessions. As Blitz noted:

> Of the 90,000 people who had been displaced from the Danube region prior to 1995, only a handful of Serb returns were facilitated by the two-way return mechanism. The majority of Serbs displaced from the Danube region left for third countries as a consequence of harassment and psychological pressure.
>
> (Blitz 2005: 367–8)

New legislation was passed under international pressure to revise its policy, yet the 'Mandatory Instructions' and the 'Programme for Return and Housing Care of Expelled Persons, Refugees, and Displaced Persons' continued to describe Serbs as persons who had 'voluntarily abandoned the Republic of Croatia' (Blitz 2005: 367–8).

In addition to obstructing the return of Serbian minorities, local government authorities (often controlled by the HDZ) actively discouraged the departure of temporary Croatian refugees from their new homes. Officially, local authorities claimed that Croats had the intention of leaving the occupied homes to return to their places of origin, but that they were unable to do so. This prevented the return of Serbs and Muslims to their occupied homes. Yet, Harvey argues that:

> There was little evidence that the HDZ genuinely supported the return of Croats to their pre-war homes; rather, incentives were offered in many cases to people to remain in their place of refuge, or to resettle in areas that had previously been inhabited by a majority of another ethnicity.
>
> (Harvey 2006: 97)

The reluctant policy of return for ethnic Serbs is of course to be put in direct relation with the enthusiastic policy of the 'return' (even though as second- or third-generation Croats in Canada, the USA or Argentina, they had never 'left') of ethnic Croats from abroad. Here, the double logic of the inclusion of ethnic Croats and the exclusion of ethnic Serbs appears with clarity: the same movement of 'return' is handled by two different policies and administrations, with two radically different objectives in mind.

The second aspect regards reconstruction and property law. The HDZ adopted a housing and reconstruction policy that actively discriminated against ethnic minorities while favouring Croats, especially those willing to resettle from abroad. The housing problem was a central question. During the war, refugees from one 'cleansed' camp begun to occupy houses left by 'cleansed' populations of the other camp, at the rhythm of the victories and defeats of the armies. An estimated 195,000 homes were destroyed during the war, and while the government conducted a policy of reconstruction (with the support of international donors) in allocating houses, it discriminated between ethnic Croat and minorities (Blitz 2005: 368). Another issue related to housing was (and remains) the question of tenancy rights to socially owned properties. Under the SFRY, many resided in houses belonging to a state-owned company or a public administration. With the switch from a socialist to a free-market economy, property was privatized. With the Law on Temporary Take-Over and Administration of Specified Property (LTTP) of 1995,[27] Croats were redefined as 'settlers' and were privileged in the adjudication over private and state-owned property. The law revoked any rights for citizens absent for more than three months, and coincided with the outbreak of the war. Thus, the provision targeted the Serbs who had fled during the war. Only in 2002 was a law passed to provide alternatives for those excluded, but no action has been taken in this direction (Blitz 2005: 368; Harvey 2006: 93; International Crisis Group 2002). Moreover, many court hearings were conducted in the absence of the Serbian owners or beneficiaries, which further dispossessed an estimated 23,700 people in 2003 alone (Human Rights Watch 2003: 34).

Finally, the judicial system blocked Serbian return and reintegration in more than just property rights. During the years following the Dayton Peace Accords, the (often unfounded) indictment of 'war crimes' under the 1996 amnesty law was used both against individual returnees and as a technique to deter further returns in specific areas. About 1,500 Croatian Serbs were indicted after the war – and many were arrested – even after they had been 'cleared' for return by the Croatian authorities. Simultaneously, Croats wanted for war crimes by the International Criminal Tribunal for Yugoslavia (ICTY) were protected by the state (Blitz 2005: 372; Harvey 2006: 93; Human Rights Watch 2004).

Following the 2001 'Knin Conclusions', however, the new left-wing majority marked a change, by passing a new Constitutional Law on National Minorities[28] (Blitz 2005: 368) to introduce improvements for the representation of minorities in the Sabor and other provisions, as well as the Law on Areas of Special State Concern[29] that reduced the discrepancy between Croats and minorities in issues of tenancy rights.

Conclusion

In the case of Croatia, post-territorial citizenship operates, through the formal and physical exclusion of the unwanted present, and the formal and institutional inclusion of co-ethnics abroad, as a form of post-territorial nationalism. The trans nationalization of ethnic nationalism is expressed through the deployment of citizenship as the main technology of inclusion/exclusion, as well as institutional arrangements and arbitrary state practices. The high point of this purely exclusionary logic was reached during the 1990s, and begins to decline with the ousting of the HDZ from power in 2000. What is later suggested as a response by the center-left coalition, with the support of the European Union (EU) and other international organizations, is not a shift to a civic understanding of citizenship, but rather a refocusing of the political sphere on the territorial borders, and the pluralist inclusion of the excluded communities, mostly the Serb minority. Yet, while the subsequent years have witnessed the introduction of increased minority rights and political representation, in parallel with a continuous attempt at reducing the weight of the diaspora in Croatian political life, the principle of inclusion of co-ethnics abroad, as well as their entitlement to participate in political life, remains unchallenged to this date, suggesting that the transnational understanding of the 'Croatian nation' has become part of the unquestioned, commonly shared assumption of Croatian political life.

For the concept of post-territorial citizenship, the Croatian case illustrates how specific transnational social and political spaces are generated through state practices – as well as the resistance to these state practices beyond the geographical confines of the territory. It brings empirical evidence to the fact that post-territorial citizenship can work as a way of channelling exclusionary politics based on religion and ethnicity. But contrary to a certain reading of post-territorial politics as apolitical, because it escapes the sphere of territorialized representative politics, the transnational space is very much a political space of contestation. What is in question, however, is whether political contestation will – as various actors such as the EU and the Social Democratic Party advocate – call into question the post-territorial character of these politics, pushing for a closure of the transnational political space, or whether post-territorial citizenship will be invested and de-ethnicized – as for example in Macedonia, where the diasporic citizenship policies have been opened to other constitutive ethnic groups – consecrating, therefore, post-territorial citizenship as a legitimate practice of citizenship.

Notes

1 Much of the research presented in this chapter has been jointly conducted with Igor Štiks in the framework of the European Union Democracy Observatory (EUDO) – see Ragazzi and Štiks (2009). I would like to thank Igor Štiks for allowing me to draw on our common work, as well as Rainer Bauböck and Wiebke Sievers for their support for the project.
2 This play on words has also been used by Cauvet (2011).
3 See Soysal (1994). For variations on the concept of post-nationalism, see Habermas and Pensky (2001) and Jacobson (1996). For a different take on post-nationalism, see the concept of 'denationalization' as developed by Bosniak (2000) and Sassen (2006).
4 'Methodological nationalism' has been developed as a concept to argue that most of the contemporary social science has, from Weber and Durkheim up until recent years, taken for granted the ideas that 'societies' are almost naturally contained by

the territorial boundaries of the state, and hence has been blind to social and political processes taking place at the interstice of these frontiers, or simply beyond them. Methodological nationalism has been introduced by Martins (1974), Smith (1979), and more recently re-discussed by Beck (2002b) and Wimmer and Glick Schiller (2003). For an in-depth study of methodological nationalism, see Chernilo (2006).

5 Hansen (2009) and Joppke (2003).

6 In most of the literature on citizenship and cosmopolitanism, post-national and post-territorial citizenship are used interchangeably. See, for example, Squire (2009), and from a standpoint critical of the concept, Chandler (2007, 2009), Pugh et al. (2007: 107) and Baker (2007). My goal here is to carve out a different understanding of the term.

7 I would like to thank Guido Tintori for highlighting the importance of this point.

8 See the history of Jewish autonomous region – Birobidhzan, in Levin (1990).

9 As Brubaker (1996) has shown.

10 *Official Gazette of Democratic Federal Yugoslavia* 64/1945. The law was confirmed and amended on 5 July 1946 (see *Official Gazette of the Federal People's Republic of Yugoslavia (FPRY)* 54/1946). The law was further amended and revised in 1947 (see *Official Gazette of the FPRY* 104/1947) and twice in 1948 (see *Official Gazette of the FPRY* 88 and 105/1948).

11 Unlike the USSR it was not printed on identification documents, but was present in other administrative registers.

12 *Official Gazette of the Republic of Croatia* 53/1991; modifications and amendments in *Official Gazette of the Republic of Croatia* 28/92.

13 A significant number of the Croatian Serbs continued to live in territory controlled by the Croatian authorities. They managed to regulate their status either smoothly (i.e. as holders of the former Croatian republican citizenship they were automatically registered into the new registries of citizens), or in some cases, with considerable difficulties. Numerous reports testify to cases of violations of their right to Croatian citizenship in the 1990s. See, for instance, reports on the issue published in Dika, Helton & Omejec 1998 and also the report on Croatia in Imeri 2006.

14 *Official Gazette of the Republic of Croatia* 68/1995, (Odluka o proglašenju Zakona o izmjenama i dopunama zakona o izborima zastupnika u Sabor Republike Hrvatske).

15 *Official Gazette of the Republic of Croatia* 116/1999.

16 Ministarstvo Povratka i useljeništva. 'Zastupnički portret: Ivan Nogalo' *Bilten ministarstva povratka i useljeništva*, no. 27, April 1998, p. 9.

17 Constitutional Law on Human Rights and freedoms and on the Rights of Ethnic and National Communities or Minorities [Ustavni zakon o ljiudskim pravima i slobodama i opravima etničkih i nacionalnih zajednica ili manjina u Republici Hrvatskoj, NN65/1991, and amendments NN 27/1992], hereinafter Constitutional Law (UZEM).

18 *Official Gazette of the Republic of Croatia* 34/1992. art. 21-51, amended text.

19 art. 10, 'Zakon o izborima zastupnika u Sabor Republike Hrvatske' [Law on the Election of the Representatives to the Sabor of the Republic of Croatia], *Official Gazette of the Republic of Croatia* 22/1992; art. 16 'Zakon o izborima zastupnika u hrvatski državni Sabor' [Law on the Election of the Representatives to the Croatian State Sabor], *Official Gazette of the Republic of Croatia* 116/1999.

20 See http://www.vlada.hr/hr/naslovnica/o_vladi_rh/prethodne_vlade_rh/3_vlada_republike_hrvatske (accessed 20 August 2009).

21 One of them, Marijan Buconjić had been convicted in the USA of assaulting a Yugoslav diplomat. See http://cases.justia.com/us-court-of-appeals/F2/581/1031/279785/ (accessed 20 August 2009).

22 Antun Babić (Australia), Domagoj Ante Petrić (Argentina), Marijan Buconjić (USA), and Nikola Vidak (Canada). Ministarstvo povratka i useljeništva "Ministar Petrović predstavio svoje pomoćnike" *Bilten ministarstva povratka i useljeništva*, no. 2, February 1997, pp. 3–6.

23 Ministarstvo povratka i useljeništva 'Govor Ministra Petrović Hrvatima Vancouvera.' *Bilten ministarstva povratka i useljeništva*, no. 1, January 1997, pp. 2–3.
24 Ibid., pp. 3–5.
25 'Croat Diaspora Meet In Zagreb – Part I', *Zajedničar*, 13 October 1993, p. 8.
26 The right to return is explicitly contained in the International Covenant on Civil and Political Rights (ICCPR), G.A. res. 2200A (XXI), 21 U.N. GAOR Supp. (No. 16) at 52, U.N. Doc. A/6316 (1966), entered into force March 23, 1976, article 12. Croatia became a party to the ICCPR in 1991(Human Rights Watch 2006:2).
27 *Official Gazette of the Republic of Croatia* 73/1995, 7/1996 and 100/1997.
28 *Official Gazette of the Republic of Croatia* 155/2002.
29 *Official Gazette of the Republic of Croatia* 26/2003.

Part III

Locating diaspora in and beyond Germany

5 Cosmopolitanism in Kazakhstan

Sociability, memory and diasporic disorder

Rita Sanders

Prologue

In Kazakhstan, in 2007, the motto of the May Day was redrafted. Instead of celebrating the internationally recognized Labour Day, as was previously done in Kazakhstan and during Soviet times, the state's President Nursultan Nazarbayev decided to proclaim the slogan 'unity of the people of Kazakhstan' (*edina naroda Kazakhstana*). Such a change in focus might be interpreted as a shift from global belonging to national introspection. In the city of Taldykorgan, where I conducted fieldwork in 2007,[1] the festivities of May Day were organized as a procession through the town, with businesses and university departments specially requested to participate.

I accompanied Viktor, a business man, some of his employees and his family on the roughly two kilometres long procession which ended in the central place in front of the *akimat* (town hall) where the *akim* (mayor) and other officials were sitting on a tribune. Every company or university department passing by was proclaimed through a loudspeaker over the sound of shouting spectators. Most of them wore peaked caps and were waving Kazakh flags. Natalya, Viktor's wife, and their three children were noticeably dressed up for the event. She told me later that they only took part in order to appeal to the *akim* and to present themselves as proactive followers. Viktor, however, also seized the occasion to chat with many people along the roadside. Viktor and his wife run two enterprises, a garage for repairing cars and a farm where they cultivate apples, berries and other crops.

Cosmopolitanism in Kazakhstan?

At first glance, a country such as Kazakhstan does not seem to be the right place for cosmopolitan attitudes and practices among its citizens. In 1991, when the state was founded, Kazakhs made up a little less than half of the population, but since then, millions of mostly Russian-speaking Kazakhstanis have emigrated to Russia, Germany and Israel. However, the country is still home to more than 100 ethnic groups. According to official records for 2009, 63 per cent of the population is Kazakh, 23 per cent is Russian, and the number of Germans is estimated at 178,000 thousand or 1.1 per cent of the total population. Thus, more than 80 per cent of the one million Germans residing in Kazakhstan in 1989 have emigrated. The state's Soviet past is still present in terms of ethnic folklore and the significance of nationality as an organizing tool (Brubaker 1996; Dave 2007; Hirsch 2005). The state's only president thus far, Nursultan Nazarbayev,[2] has, furthermore, pursued nationalizing policies which favour ethnic Kazakhs by promoting the Kazakh language at the expense of Russian as the previous lingua franca.[3] For instance, Kazakh primary and secondary education as well as university education has been promoted nationwide, and the 1997 law on languages also stipulated that at least 50 per cent of all television and broadcasting should be in Kazakh (Dave 2007: 102; Schatz 2000: 494).

While previously cosmopolitanism has been associated with a cosmopolitan world class of, for instance, diplomats, in recent years the notion has also been applied to explore the experiences of people who are locally rooted or do not belong to the 'West'. In this vein, Featherstone (2002: 2) asks: 'What equivalent forms of cosmopolitan experiences, practices, representations and carrier groups developed, for example, in China, Japan, India and the Islamic world?' The Western ideal, 'which derives from the Kantian tradition and entails some notion of a *polis* extending around the globe' (ibid.: 3), he concludes, should, thus, be seen in the plural. This recent multiplicity of cosmopolitanisms has been termed, for instance, 'vernacular cosmopolitanism' (Werbner 2006) or 'socialist cosmopolitanism' (Hüwelmeier 2011). What social scientists investigate in studies on cosmopolitan attitudes and practices is usually someone's openness towards the 'other', who is conceived of as different, and, secondly, her or his multiple attachments and ability to cope with diverse cultural settings.

But the application of cosmopolitanism to labour migrants, shopkeepers and pilgrims, which implicates empirical anchoring of the philosophical ideal, has also resulted in a critique of the concept itself. First, it turned out that feelings of national and ethnic belonging do not circumvent cosmopolitan attitudes and practices. Thus, from an anthropological point of view, cosmopolitanism is not the longed-for overcoming of nationalism, as has often been assumed (cf. Beck and Sznaider 2006). Instead, the dual poles of nation and cosmos have become reconciled by coining the notion of 'rooted cosmopolitanism' (Cohen 1992), and Glick Schiller *et al.* explicate (2011: 400) that 'rootedness and openness cannot be seen in oppositional terms but constitute aspects of the creativity through which migrants build homes and sacred spaces in a new environment and within transnational networks'.

Secondly, recent studies have questioned the connection of mobility with cosmopolitanism, which has been central to writings on the subject (Beck and Sznaider 2006; Hannerz 1990). The experience of spatial movement is no longer seen as a prerequisite for openness and multiple attachments (Marsden 2008; Notar 2008) and being mobile does not necessarily alter one's attitude towards people whom she perceives to be different (Halemba 2011). Instead, anthropological research studies are concerned with individual variation, diverse local practices and 'situated cosmopolitan openness' (Glick Schiller *et al.* 2011: 400).

The empirical complication of a project which is intrinsically normative is, thus, under way. This chapter seeks to add to the existing body of literature on empirical cosmopolitanism by explicating another case study that brings up further questions and concerns. The people I depict here are usually referred to as diaspora and most of them indeed identify themselves as Germans who live in Kazakhstan, although their relationship towards their presumed 'homeland' is anything but easy. Gorbachev's policies of glasnost allowed Soviet Germans to emigrate and to settle in Germany. During the late 1980s and, in particular, after the collapse of the Soviet Union, a huge flow of more than 800,000 Germans left Kazakhstan in order to start a new life in Germany. This has resulted in the vast majority of remaining Kazakhstani Germans having relatives and friends living in Germany, but often with disappointment at their new life in their 'historic homeland'.

This chapter, however, is concerned with those who did not migrate, even though this meant that they had to separate from many of their relatives, friends, colleagues and neighbours. Therefore, Kazakhstani Germans do not appear to be the obvious candidates for cosmopolitan behaviour. But I argue that their staying is exactly what allows them to keep a cosmopolitan attitude, since they have not lost the cosmos in which they can be cosmopolitan. This brings me to question the equation of cosmos with a world community, which does not empirically hold. On the contrary, I emphasize the role of states and their policies, which also implies a critique of the assumption of viewing cosmopolitanism merely as a set of personal characteristics. In this vein, I look at places and occasions for sociability, which are partly organized by state officials and which seem to foster engagement with the other. Furthermore, I stress the impact of experienced hospitality on cosmopolitanism, and I argue that such experiences might exert enormous power when they have entered a community's cultural memory.

Interethnic sociability

The state of Kazakhstan is generally understood as taking up the Soviet policies of internationalism and interethnic harmony while equally promoting ethnic Kazakhs. In this ambivalent endeavour, politicians seek to strengthen a supra-ethnic Kazakhstani identity (Cummings 2006; Dave 2007; Diener 2004; Holm-Hansen 1999; Oka 2006). In the following, I elaborate the concept of internationalism in Soviet times and present-day Kazakhstan by looking at the policies' output in terms of practices of sociability and people's attitudes towards others.

Though the concepts of nations and nationalities have no place in Marxism, since they are an essentially bourgeois phenomenon, they received a key role in organizing and controlling the Soviet people (Hirsch 2005; Martin 1998, 2001; Pilkington and Popov 2008; Slezkine 1994). However, at the same time, nations and nationalities were largely depoliticized identities and Soviet internationalism is usually characterized as formalism resting on cultural presentations like dancing or singing, which are performed by ethnically defined groups (cf. Humphrey 2004: 146; Hirsch 2005: 269–71).

Humphrey (2004) has investigated the interrelation of internationalism and cosmopolitanism in the Soviet Union. She underlines that internationalism requires the idea of different nations and, thus, is to be distinguished from a concept which builds on their overcoming. Referring to Foucault, she states (ibid.: 142f) that '[…] an idea of "infinite openness" is inimical to any system of thought resting on the value of emplacement and the relations between placed entities. In this respect, nationalism and internationalism align with one another, and neither of them have any room for cosmopolitanism'. The Soviet version of cosmopolitanism (*kosmopolitizm*) was usually combined with the adjective 'rootless' (*bezrodniy*), which was a severe accusation that sent people to prison during Stalinism (ibid.: 141–3). But, paradoxically, Soviet internationalism has fostered cosmopolitan thinking and acting, or as Humphrey (2004: 146) puts it: 'Yet for all that, in actuality this was also a moral universe of comradeship', which she explains by elaborating the following: 'Some cities, which had been ethnically mixed from the start and gained further contingents from the Soviet development policies, became truly cosmopolitan spaces. Almost despite itself, internationalism – because it denied conflict and encouraged common values – enabled an unacknowledged cosmopolitanism to flourish' (ibid.).

Thus, in effect, the Soviet Union might be interpreted as a cosmopolitan project which allowed or expected its citizens to behave with open-mindedness and tolerance towards others, but which, at the same time severely punished such relations with the 'moral outside' (Humphrey 2004: 147). The cosmos for cosmopolitan practices was therefore clearly marked. Cosmopolitan acting was, furthermore, seen as a necessary prerequisite for building socialism and a unified Soviet identity and even deepened the border to the outside world. The fact that the Soviet cosmos was clearly defined and its development solely perceived as a project of science might be connected to the condition of cosmopolitanism as such, as Latour (2004: 453) elaborates:

> […] whenever cosmopolitanism has been tried out, from Alexandria to the United Nations, it has been during the great periods of complete confidence in the ability of reason and, later, science to know *the one* cosmos whose existence and solid certainty could then prop up all efforts to build the world metropolis of which we are all too happy to be citizens. The problem we face now is that it's precisely this 'one cosmos', what I call *mononaturalism*, that has disappeared.

The question, thus, is what has happened to the Soviet cosmos? Is it still to be found in Kazakhstan? I argue largely yes. To begin with, cultural variety and interethnic accord are on many occasions publicly displayed. A representative of a German minority centre[4] explicates such festivities as follows:

> I think things could not be better [he is referring to interethnic relations] because even we as representatives [of the minority centres of Taldykorgan] go to the annual meetings of the Assembly of the People of Kazakhstan,

which take place either in Astana or in Almaty. There, our President Nursultan Nazarbayev takes part. And the relationship between the representatives of the national minority organizations, well, I think, it could not be better. One for all, and all for one. [...]

Well, in general festivities proceed in every cultural-national centre. We stand shoulder to shoulder. And every centre has its own songs. It's all a huge fair. One of the festivities is May Day. It takes place every year. The *akim* (mayor) of the region and the *akim* of the town take part and congratulate us. Afterwards, a public concert takes place. It's a big concert on a big stage, a public stage. And every cultural centre – national cultural centre – which takes part shows its character sketch. It's a fairly interesting spectacle. You will still be here on May Day. You must necessarily visit it. It's impossible to tell about it in all detail – it's better to see it once. This is how it is.

(Igor,[5] representative of the German minority centre [*nemetskii dom*[6]]; February 2007)

Igor describes the festivities of May Day as they were previously celebrated, with booths of the minority centres and other associations or companies exhibiting a variety of products or their 'culture' under the heading of Labour Day. He had no problems when he heard of the changes to May Day some months after our interview, saying he appreciated the new motto of the 'unity of the people of Kazakhstan' (*edina naroda Kazakhstana*). The envisioned unity might be a delayed output of the Soviets' intention to merge the diverse nationalities and nations into one single socialist people. During Soviet times this project became ever more postponed (Hirsch 2005: 319f). However, in current-day Kazakhstan it still seems to be pursued and the fostering of an all-encompassing Kazakhstani identity is to be seen in that light.

The state-sponsored creation of a Kazakhstani identity is usually criticized as being a disguised form of ethno-national policies and is widely questioned as an attractive mode for non-Kazakhs (Cummings 2006: 191; Dave 2007: 135; Holm-Hansen 1999: 162).[7] Certainly, state policies have mostly aimed at improving the situation for ethnic Kazakhs. This was done by enhancing the status of the Kazakh language, which in turn ameliorated job opportunities for Kazakh speakers and resulted in more privileges for ethnic Kazakhs (Davis and Sabol 1998; Peyrouse 2007: 486).

Though I do not intend to doubt the effects of such nationalizing policies,[8] I wish to make other points here. First, it is absurd to assume that Russians or, better yet, Russian-speaking minorities will be culturally suppressed: the policies hardly affect the willingness of Russians to learn Kazakh. Secondly, non-Russian minorities in particular are encouraged to develop their cultural identity, which is, certainly also done with the intention of weakening the most populous Russian minority. Germans, for example, mostly interpret such policies as aiming to improve their situation. Finally, and regarding the festivities I attended during my year in Taldykorgan, all apart from one – namely the revised May Day – included exhibitions of minority booths representing their 'culture'. Therefore, ethnic identities in general do not seem to be under threat. However, by boosting 'Kazakhstani identity', state officials seek to reinvent the Soviet cosmos on a national level. The socialist people can no longer be dreamed of, which is why state officials turned their aspiration to the 'people of Kazakhstan'. The values of internationalism which once governed the Union's policies are still the same, but the cosmos they refer to today is much smaller.

Thus, Kazakhstan's state policies might be coined as 'rooted cosmopolitanism', since feelings of ethnic belonging are assumed to be a prerequisite for ethnic accord, and thus, for cosmopolitanisms. But what then of the Kazakhstanis themselves: Do they perceive themselves as such? As already mentioned above, a Kazakhstani identity is usually challenged for meaning by those living in the country, and indeed, people rarely refer to themselves as 'Kazakhstani' [*Kazakhstantsyi*]. However, the fact that the notion is not very common does not necessarily mean that the concept is not shared. If one asks people 'What is your nationality?' most of them will answer by pointing out, for instance, 'Russian', or 'Uzbek', but many of the Germans I interviewed added, 'German, but my homeland is Kazakhstan', and they attach cosmopolitan values to this homeland. So people often explicitly say that they love Kazakhstan for its interethnic harmony and open-minded attitude toward others. A statement by the 22-year-old Sonya, who regularly visits the gatherings of young people in the German minority centre, illustrates this view of cosmopolitanism:

> My circle of friends is ample and multinational. Recently, we were sitting together and counting: We probably have six to seven different nationalities. We are all friends and often come together. And in such an association this is not recognizable. In every case, whenever you meet someone, the boundary becomes blurred, for instance, that he is of another nationality. I think this is good. If we spoke like this 'he is Russian, this is a German' there would nevertheless be a kind of barrier in our communication. But if this is absent, I think, communicating, working, and living are improved – when there are no barriers, especially in regard to nationality. Well, it's like this.
>
> (Sonya, a member of the youth club of the German minority centre [*nemetskii dom*]; June 2007)

Sonya's statement is a true account of a cosmopolitan 'world view', or better yet, a 'Kazakhstan view'. People know their roots, but ethnic differences of any kind should not imply barriers, neither in regard to all sorts of encounters in general nor to friendship in particular. A comment like Sonya's is not exceptional, but widely shared among the youth club's members of the German minority centre (as very likely among most members of all other ethnic minority centres). This might be astonishing since one assumes minority centres to act first of all for the rights of the particular ethnic group they seek to represent. However, one must keep in mind that the authoritarian state of Kazakhstan does not allow for any political grassroots movements to develop.

All contemporary political associations are to some degree controlled by the state, and the minority centres themselves are pooled under the Assembly of the People of Kazakhstan (*Assembleya Naroda Kazakhstana*), which is headed by the President himself.[9] Therefore, minority centres are also chiefly concerned with propagating the 'cosmopolitan Kazakhstan view'. This, however, does not mean that those people who work and visit the minority centres do not really live and feel the cosmopolitan spirit. As far as I can judge, many individuals even emphasize this as a crucial part of their work or as part of what a minority centre stands for. The quasi state task of guaranteeing interethnic harmony is partly fulfilled by the minority centres and their efforts at organizing all kinds of festivities, be it the town's day, *Nauryz* (the 'Kazakh New Year') or the state's Independence Day. Such events are often characterized by their purely formalist nature, from which it is often concluded that their impact on people's attitudes and practices is mild. Dave (2007: 130), for instance, explicates that 'consistent with the Soviet socialist formulation, nationality is imagined as "cultural", that is, a non-political community. Within this framework, the state allows, in fact encourages, official national-cultural centres to engage in "cultural" and "ethnographic" activities', thus, 'culture is constructed in a folkloristic sense, with fixed meaning and insignias, devoid of a subjective identity dimension' (ibid.). To be sure, it is certainly right that most people are aware of the merely folkloristic character of such events and do not believe, for instance, that Kazakhstani Germans dance polka in 'real life', as is presented at most of the festivities in Taldykorgan. Thus, the declared aim of such expositions, namely to inform people about cultural habits, is surely not accomplished. However, people gather, meet and celebrate together, for whatever reason that might be.

The concept of sociability was only very recently brought in touch with cosmopolitanism by Glick Schiller *et al.* (2011). The authors investigated the situated nature of cosmopolitan practices by also looking at particular places: 'These places [of cosmopolitan sociability] are created through people's meetings, encounters, civic communication and coexistence' (ibid.: 403). Sociability (*Geselligkeit)* as a concept was formulated by Schleiermacher during the Enlightenment in the late eighteenth century. Encounters of sociability are defined by their open and non-instrumental character, which distinguishes them from deliberately acting at the workplace or at home within the family (Schleiermacher 1984 [1800/1810]). Sociability is – as well as cosmopolitanism – a normative project which aims to realize that every participant of a parlour becomes familiar even with the most foreign minds and circumstances in order to be able to feel close to them (ibid.: 41). This objective fits well within Kantian cosmopolitan projects. In order to foster moments of sociability, people should meet at particular places where they ideally follow certain rules of communication. For instance, a topic is chosen with the aim of not excluding any of the participants of the particular group from sitting and discussing together. Such gatherings can only ensure that people of different professions and classes come together. Furthermore, everybody feels free to utter their personal opinion and should not hide their evaluation in order to guarantee that people can really learn from each other's point of view and broaden their perspective. Only in that way can a society, as a whole, develop.

Encounters of sociability are, to my view, a prerequisite for cosmopolitan practices and attitudes, though the rules of sociability formulated by Schleiermacher cannot be deployed directly to a club in an authoritarian state like Kazakhstan. But the ideal fits well within the objective of the youth club in the German minority centre where teenagers of various nationalities meet on a weekly basis. Usually they just hang out, dance in the building's basement, chat and make friends, or wish to date someone. However, such gatherings mostly have an official framing when the head of the youth club welcomes everybody and – when it is the case – explains the day's programme. For instance, all official state festivities are solemnized in the club. On these occasions, the participating teenagers might, for instance, prepare a sketch by working together in small groups while the head of the youth club ensures that all of the participants' national traditions are equally represented in the final programme. In terms of sociability, it is less the final show which is decisive, but the fact that the participating teenagers regard their weekly gatherings as a serious endeavour in which they learn about each other's point of views within an organized, but ultimately non-instrumental framework.

The same holds true for festivities in Kazakhstan in general, which all seem to have a normative flavour, often a cosmopolitan one. The Kazakhstani version of cosmopolitanism is, thus, realized through sociability. Festivities and state organizations like the minority centres create places where different people can meet, spend some time, and become familiar with 'foreign minds'. Also in day-to-day life, people attach significance to invitations and festivities for establishing good neighbourly relations with people from other ethnic backgrounds. The following quote from a 47-year-old bank accountant illustrates this:

> However, the people [*narod*] who live here comprise different nationalities. Here, in principle, it's no one's business which nationality you belong to. People notice what kind of man you are, how you work, and how you behave towards others. We even live here as neighbours. Fatima lived here. Her mother comes from the Caucasus, her father was a Kazakh. Now she has emigrated. A German family lived here, then after them Russians and now Kazakhs. Well, in general, there are no problems. They respect our rules. For instance, they accept our invitations to our parties. They celebrate with us, and we visit them.
>
> (Bank accountant, 47 years old; April 2007)

Kazakhstan's interethnic harmony is closely connected to moments of sociability. People can thereby resort to practices of feasting in settings with diverse nationalities, which they frequently did during Soviet times and after. Consequently, the state-sponsored festivities have entered the day-to-day practices of ordinary citizens and the value of interethnic accord is widely shared.

Memory of hospitality

Cosmopolitan attitudes are not only widespread because of interethnic festivities but also because of people's experiences of hospitality, which are stored in their memories. In the following two sections, I focus on those people who themselves or whose ancestors were deported, mostly from the Volga Republic or the Black Sea Coast, to either Siberia or contemporary Kazakhstan, because Soviet Germans were feared to collaborate with the German Reich after the attack of the *Wehrmacht* on the USSR in 1941. Immigrants from the discrete German states had mostly come to settle in the Russian empire during the late eighteenth century. In present-day Kazakhstan, some of their descendants still refer to themselves as Germans. Their cultural memory is marked by movement and by the pride of getting through various, often hard, circumstances. The situation after deportation is, however, particularly tragic. Several hundred thousand deportees died, mostly of starvation during the winter of 1941–2. Germans in contemporary Kazakhstan tend not to stress this fact, but instead stress the fact that the newcomers, who were officially titled 'the people's enemy' (*vrag naroda*) were generously helped by the local Kazakh population.

Older experiences of mobility are rarely pivotal in research on cosmopolitanism, which might be linked to the concept's recent prominence (in the social sciences) and the fact that cosmopolitan practices are implicitly connected to the 'age of migration' (Castles and Miller 1993). An exception is Marsden (2008), who explores the linkage of old experiences of mobility with present-day engagement with globalizing processes in the border area between Pakistan and Afghanistan. He concludes that 'cosmopolitan practices are neither the one-dimensional products of civil war nor the straightforward outcomes of global migration. They bear the imprint, rather, of older cultural influences – Persianate practices of travel, for example' (ibid.: 240).

The memory of the hospitable Kazakhs who equally suffered during the war, but who, nonetheless, shared their bread with the deportees is still alive today and told in many personally modified versions. It seems that parents and grandparents emphasized this part of their family histories and, furthermore, that the younger generations find those narratives particularly momentous. The statement of Vera, who is 62 years old, exemplifies this:

> The Kazakh people [*narod*] have always behaved well towards us. Between our peoples, in principle, there have never been any disputes. For instance, my parents have always told me that the majority of their generation, who came during those years [during the Second World War], in principle, only survived and did not perish because of the fact that the Kazakh people are very benevolent. Very many people helped. They [the deportees] were abandoned, basically, without anything. Therefore, today, I do not feel uncomfortable with the fact that I'm German. The only thing which is a little bit uncomfortable is the fact that we have not learned the Kazakh language very well. This we regret today.
>
> (Vera, 62 years old; February 2007)

Vera's remarks are not exceptional in that she underlines that Kazakhs saved the lives of those people who were forcibly settled in the Kazakh Republic in the course of the Second World War. Therefore, in their memories, Kazakhstani Germans are deeply indebted to Kazakhs' generosity. The view of Kazakhs as extraordinarily hospitable people is generally widespread and also endorsed by the Kazakhs themselves and in state discourse on their history and self-conception. Thus, Kazakh hospitality is also a stereotypical characteristic which supports the deportees' memory.

The concept of hospitality has its roots in the same orbit of enlightenment as do cosmopolitanism and sociability, and like the others, it is a normative project which is developed by the formulation of universalist laws, as in Kant's famous 'Perpetual Peace' (1795). The concept has entered the social sciences by the growing recognition of mobility and immigration practices and by the late work of Derrida who explores 'the idea of a receptiveness to the arrival of an other which breaks through any prior assignment of roles, duties, conventions' (Dikec *et al.* 2009: 3). For Dikec (2002), hospitality cannot be captured by defining a set of rules; instead, it is relational and connected to recognition 'which is stronger than tolerance, with its more passive stance of mutual non-interference. It implies the act of engaging with and reverence for the stranger, in contrast to much of the current practice towards guest workers, migrants and refugees we find around the world' (Featherstone 2002: 7).

Thus, one may find that hospitality is the demanding version of cosmopolitanism because of its emphasis on practicing engagement with the other. Or, in other words, it is the assignment of practices which entail the remedy of how to implement the rather amorphous cosmopolitanism. At least in the case of the deportees' descendants in present-day Kazakhstan, it is very likely that the transmitted experiences of hospitality have contributed to cosmopolitan attitudes. To be sure, the Kazakhs have to be called cosmopolitans in the first place, but the point here is that experienced cosmopolitanism – and in particular hospitality – might be retained in people's memories and still be pivotal, even if the situation has changed considerably. The earlier quote by Vera indicates exactly this. Kazakhstani Germans – as all other Russian-speaking Kazakhstanis – have faced difficulties since the implementation of Kazakh as a single state language. For example, at present, it is difficult for them to find employment in the state sector.[10] Nonetheless, Vera expresses that she does not feel uncomfortable as a German in present-day Kazakhstan, which she explains by the experienced hospitality of her parents. She only utters regret about the fact that she did not learn Kazakh well enough. Therefore, it is the memory of hospitality, which entails that she feels associated with the new rulers of her homeland. This is not as trivial as it might seem at first sight because usually (only) Russian speakers still feel superior to the Kazakhs once classified as underdeveloped (cf. Peyrouse 2008: 109). It is certainly also against this background that Kazakhstan's policies of interethnic harmony have to be seen.

Kazakhstani Germans, however, often strive to establish a special bond between themselves and Kazakhs. This is also done by highlighting that they speak Kazakh, as do many of those who were deported during childhood or born shortly after the Second World War and raised in Kazakh families, as the following narrative by Ludmilla exemplifies:

> With the Kazakhs we lived together in great, great harmony. There were almost no other nationalities, only a few Russians. Our neighbours were all Kazakhs. Grandma Roza [a Kazakh woman] raised me. My mother went to work in the fields – I was about four years old – and brought me to this grandmother. Therefore, I speak Kazakh. Look, she has raised me. I was more together with her. My mother came home late at night and went to work in the mornings. Well, I'm very proud of – I don't know why – that I know Kazakh. Even at school I sang in Kazakh. The others couldn't speak [in Kazakh] like I did.
>
> (Ludmilla, 72 years old; February 2007)

Those children were forced to engage with the other's culture and most of them – like Ludmilla – are today proud of having obtained the competence of getting along in this foreign culture. Featherstone (2002: 13) remarks that 'the prospects for cosmopolitanism depend upon the maintenance of a certain degree of "world openness", a capacity to embrace the culture of the other or receive the other's culture unconditionally, as we noted above in the discussion of hospitality.' Kazakhstani Germans were during their history constrained to embrace and receive the other's culture. But this experience is treasured with good feelings and might have fostered cosmopolitan practices and attitudes.

Diasporic disorder

Kazakhstani Germans' memory is ethnically loaded, which crops up explicitly in narratives on forced resettlement during the Second World War, the subsequent loss of German language skills and recent transnational ties to a 'historic homeland' [*istoricheskaya rodina*]. But up to the 1980s, relations to and information about Germany was extremely rare and a definition of being German was largely detached from the remote homeland which, however, had given people their identity by name, and had brought immense misery to them during the two world wars and thereafter. Nonetheless, some Kazakhstani Germans felt, and still feel, that they ultimately belong to a 'historic homeland', which has been imagined as a place of shelter and prosperity. Thus, diasporic sentiments have also contributed to the significant outmigration of Kazakhstani Germans once the opportunity was given.

The concept of diaspora has only recently been investigated against the background of cosmopolitan thinking and acting (Ali 2003; Appiah 2005; Darieva 2011; Falzon 2003; Ong 1998; Ziemer 2009). To stick to primordial ties of common ancestry is probably simply too remote from a concept stemming from the idea of world citizenship. However, next to the classic approach to diaspora (Cohen 1997; Safran 1991), which is oriented towards the Jewish diaspora by analysing the diasporic key conditions – namely, dispersal, attachment to a historic homeland and victimhood – some counter-thinking on diaspora has been developed (Clifford 1994; Hall 1990). This second approach interprets diaspora as a state of double consciousness and hybridity. However, it contributes rather to the broader field of studies on globalization and mobility and does not hold for analysing people's attachment to a remote homeland (cf. Anthias 1998). One of the exceptions in the field of diaspora and cosmopolitanism is Darieva (2011), who investigates the Armenian diaspora and their global ecological activities. She can show that although people maintain diasporic narratives and identities, they might strive for cosmopolitan practices. This clearly supports the broader assumption of 'rooted cosmopolitanism'.

In the two previous sections, I have argued for interpreting Kazakhstani Germans and other Kazakhstanis as cosmopolitan thinkers and actors. However, I have made explicit that their cosmopolitanism is restricted to the space of the previous USSR, which may even be restricted to that of present-day Kazakhstan. In this section, I explore the boundaries of Kazakhstanis' neat cosmos and disclose their relations and attachments, which lead beyond it. In doing so, I first return to Viktor, who was introduced at the beginning of this chapter. After the May Day parade, I accompanied Viktor and his family home. They had invited some friends and we were sitting around a well-set table when Viktor gave the evening's first toast. He praised the beauty of the country and its tolerant people and highlighted that Kazakhstan is his homeland. His – according to their self-description – Kazakhstani Russian and German guests encouraged his assertions by shaking their heads and by verbally expressing their approval. The following discussion touched on the issue of migration, both to Germany and Russia. Viktor, articulately, made clear that Germany would not be an option for him. In particular he disliked the bad food, the obstructed and polluted environment and the boring people who are not capable of drinking vodka. Instead, he expressed his admiration for China and its clever people and vivid cities where he likes to travel. Though I had the impression that not everybody fully agreed with Viktor's statements, most of his guests shared his general point and one man added the most common narrative about Germany's Germans: namely, that one cannot just come and visit someone, but instead has to agree upon a fixed appointment beforehand.

Viktor has been to Germany many times. In particular, during the early 1990s, he engaged in buying second-hand cars on German car markets, and selling them in Russia and Kazakhstan. But he – as did many of his friends – also visited his relatives in Germany. The situation of German immigrants from the former Soviet Union is generally described as complicated (e.g. Dietz 2006; Römhild 1998, 2007; cf. Pichler and Schmidtke 2004: 61). False expectations of life in Germany and difficulties in the German labour market are mostly referred to. Those who came with the vision of their historic homeland are particularly disappointed, because people in Germany usually do not accept the newcomers' German identity, but instead identify all Russian speakers as Russian, which is, furthermore, often not well-intended. Thus, Kazakhstani Germans have frequently not encountered cosmopolitan practices like they were used to in Kazakhstan and those experiences of discrimination and exclusion have been transmitted to those who stayed behind in Kazakhstan, and, sometimes, the relatives' narratives even serve to stop further migration.

Kazakhstani Germans are today much more aware of the fact that their practices of internationalism cannot be easily translated into another context, but are restricted to the world of those with whom they grew up. For some this really came as a surprise, since they thought that they are particularly experienced in internationalism and well prepared for living abroad. But interethnic sociability and hospitality are not globally shared, which is so well expressed in the narrative on the necessity of making an appointment. Thus, in some cases, people might even lose their cosmopolitan attitudes and practices because of having migrated to another place. This implies that cosmopolitanism is not a kind of essence that some people have while others do not, but that cosmopolitan practices depend on each situation. A number of similar situations might, then, form particular attitudes on cosmopolitanism. From this it follows that situations of potential cosmopolitanism beyond mobility have to be taken into account. Moreover, one should consider which particular cosmos people refer to when talking about their cosmopolitanism, since it seems very unlikely that they really mean the entire world, with all its people, regardless of which experiences they share with them.

Conclusion

Kazakhstan is one of the Soviet Union's successor states. The desire of Kazakhstan's President to strengthen the state's Kazakh appearance and identity is ambivalently amended by pursuing the Soviet policies of internationalism. Although the notion of interethnic harmony stands conceptually in contradiction to cosmopolitanism, they appear to be closer to one another if one considers people's attitudes and practices, and if one acknowledges 'rooted cosmopolitanism'.

The old Soviet and the new Kazakhstani version of cosmopolitanism essentially operate by providing moments of sociability. In contemporary Kazakhstan a vast number of state-sponsored minority centres are concerned with organizing festivities that aim to bring people 'of different roots' together to inform all involved about the other's traditions and customs. Though such events are usually described as pure formalism, they create places and instances of sociability, while celebrating together has become an indicator for people to describe good neighbourly relations.

Most Kazakhstani Germans are the descendents of deportees or were themselves deported to Kazakhstan during the Second World War. The tragedy of losing one's home is often counterbalanced by a narrative on Kazakhstani hospitality, which is often said to have saved lives. Many of the deportees were hosted by Kazakh families, who 'shared their bread' with the newcomers. This experience of hospitality has become part of Kazakhstani Germans' memories of the past and is transmitted from generation to generation. The ancestors' narratives have added to cosmopolitan practices and attitudes in present-day Kazakhstan.

Since the late 1980s, most Kazakhstani Germans have migrated to Germany. In their 'historic homeland', however, many have difficulties settling down. They find it hard to make their living and are often identified as foreign and unwelcome Russians. Or, in other words, they have had to realize that their view of cosmopolitanism is not shared in their new homeland. This message has also been transmitted to those who stayed behind in Kazakhstan. They are now aware of the fact that the cosmos of their cosmopolitanism is restricted to the sphere of the former Soviet Union, and maybe even to Kazakhstan. Consequently, many of those who praise interethnic accord now exclude Germans in Germany.

Cosmopolitanism, sociability and hospitality are all normative projects that have their roots in the Enlightenment and have only recently been applied by social scientists. The ideal of the creators of these concepts is rarely to be found in 'real life', and if ever, then only in particular moments of encounters between people. The experience of mobility does not, as others have already demonstrated, say much about someone's cosmopolitan attitudes, as people may behave very differently when confronted with new situations. Cosmopolitanism is not a characteristic of an individual person. On the contrary, moments of cosmopolitanism depend on all parts and the particular situation, or, more generally speaking, the availability of sociability. But experiences of previously experienced hospitality might be stored in one's own and one's children's' memories and be the basis for engaging with others.

Notes

1 This chapter is based on 12 months of ethnographic fieldwork, which I conducted mainly in Taldykorgan in 2006 and 2007. Taldykorgan is a city of about 135,000 inhabitants situated roughly 300 kilometres northeast of Almaty. Along with participant observation, I conducted live story interviews with about 60 respondents and semi-structured interviews with about 40 respondents. Furthermore, I collected genealogies from 40 people and ego-centred social networks of 30 people. Interviewees were mainly Germans, Kazakhs and Russians. They all belonged to different age groups and around half the respondents were female. My project was funded by the Max Planck Institute for Social Anthropology in Halle, Germany.

2 In September 1989, Nursultan Nazarbayev became chairman of the Supreme Soviet of the Kazakh SSR. Nazarbayev converted his chairmanship of the Supreme Soviet to the post of president, confirmed by parliamentary elections in March 1990 (Cummings 2005: 22). Since then President Nazarbayev has been 'officially' re-elected several times.

3 A Soviet law on languages, passed in 1989, laid the foundation stone for Kazakhstan's post-Soviet legislation. Here, for the first time, union and autonomous republics were permitted to establish 'their language' as the (single) 'local official language'. Russian was subsequently given the status of the 'language of interethnic communication' (*yazyk mezhnatsionalnogo obsheniya*) by the Kazakh SSR. In Kazakhstan's first constitution in 1993 as well as in its second passed in 1995, and in the law on

languages passed in 1997, the status of Kazakh as the single state language was confirmed (Fierman 1998: 176–80).

4 Most ethnic minority groups run official cultural centres which are integrated in the Assembly of the People of Kazakhstan headed by the state's president.

5 The interviews were all conducted in Russian by me. All interviews were taped, transcribed by me or an assistant, and translated by me. All names have been made anonymous.

6 *Nemetskii dom* means literally 'German House'. The official name *nemetskoe kulturnoe tsentre* (German cultural centre) is not particularly popular.

7 The process of nation-building has involved nationalizing policies in all of Soviet Union's successor states. The consequences for Russian-speaking minorities and their diverse responses have been intensively investigated (e.g. Barrington *et al.* 2003; Brubaker 1996; Chinn and Kaiser 1996; Jubulis 2001; Laitin 1998; Poppe and Hagendoorn 2001; Tishkov 1997). In comparison to the other Soviet successor states, it is often stated that assimilation to the titular nationality is most unlikely in Kazakhstan. Policies of nation-building, like for instance language policies, are, however, mostly evaluated as balanced. In this context, it is also hinted that the Kazakh parliament was the last republic legislature to declare its independence from Moscow in December 1991.

8 Next to the law on languages, other nationalizing policies include the reconstruction of regional *(oblast)* boundaries in the north during 1994–7 in which 'all Russian-dominated regions were merged with the neighbouring Kazakh-dominated regions' (Dave 2007: 122). Also, the capital's shift to Akmola, present-day Astana, in 1998, is seen as a measurement to attain 'a greater vigilance over the Russian-dominated regions, and to secure the loyalty of the Russified Kazakhs in these territories' (ibid.: 123).

9 The Assembly of the People of Kazakhstan was established in 1995, with the official aim to 'strengthen public stability and interethnic accord' (Oka 2006: 367). As a President's consultative body, it unites pro-regime ethnic movements, consisting of over 300 representatives of various ethnic groups and has branches at the oblast level (Dave 2007: 131). Furthermore, many nationalities maintain their 'national-cultural centre', which nominates part of the delegates to the Assembly while the President nominates others (ibid.). For Cummings (2006: 187), both Assembly and national-cultural centre are 'largely symbolic outfits' which lack any legislative power or political influence.

10 According to Peyrouse (2007: 486) '80 per cent of those working in the administration and academic circles are ethnic Kazakhs', but only 63 per cent of the total population.

6 Migration memoirs and narratives of Polish migrants in Berlin

Dorota Praszałowicz

From the turn of the nineteenth century until today, Berlin has been one of the most important centres of the Polish diaspora. Analysing the memoirs of Poles in Berlin, this chapter explores the narratives of two different groups of Polish migrants. First, this chapter examines the narratives of Polish migrants who are part of the so-called 'old' migrations – those who migrated to Berlin from the mid-nineteenth century to the 1930s. Secondly, it analyses the narratives of Polish migrants who are considered 'new' migrants – those who migrated to Berlin after the Second World War. As will be shown, although both groups belong to the Polish diaspora, they articulate their diasporic belonging to different degrees. Both groups are well-versed in living in two different cultures and positively engage with their host society. Nevertheless, while the 'old' migrants' narratives are rooted in 'ethnic absolutism', the 'new' Polish migrants seem to be practicing cosmopolitanism in their everyday lives.

Two different historical contexts (old and new) produced different collective memories, which are reflected in the migrants' narratives. The old narratives were shaped within the context of rising nationalism all over Europe as well as within the context of the Polish national struggle for survival. At the time of the 'old' migrations, 'the partitions of Poland in the eighteenth century [1772, 1793, 1795] gave the country an "essential" identity as "the Christ among nations", crucified and re-crucified by foreign oppression' (Novick 2001: 4). This identity was an inherent part of Polish collective memory in the past. Contrary to the old, the recent Polish migrants distance themselves from the 'ethnic absolutism' and indulge in the search for a new identity. Their collective memory and their narratives are woven from a fabric which has nothing to do with the Polish martyrdom and victimhood of the past; instead, it is triggered by globalization, pluralism and mobility.

Diaspora is defined here in a very broad way as a group of people who left their home country and maintain a strong collective identity (Cohen 1997; Weinar 2008). Diasporic identity is negotiated and redefined in the host country and plays a crucial role in migrants' lives. As Cohen (1997: 3) notes: 'all diasporic communities… acknowledge that "the old country"… always has some claim on their loyalty or emotions'. This very sense of belonging informs the narratives of these migrants. In the current age of globalization, diaspora is defined by interconnectivity and adaptability, allegiances to community, locality and nation. Although there is no doubt that notions of 'homeland' and 'ethnic community' have been accorded a particular primacy in the diasporic imagination, the concept of diaspora does not allow enough flexibility to properly account for the everyday experiences of Polish migrants in Berlin. While members of the Polish diaspora in Berlin display attachment to their homeland and ethnic communities, they do so in different ways and to different degrees. In this way, occasionally, they draw on cosmopolitanism as an identity resource, which does not imply an absence of belonging, but rather the possibility of belonging to more than one ethnic and cultural locality simultaneously (Werbner 1999). In this context, cosmopolitanism is seen as a practice to positively engage with 'the otherness of the other' and the oneness of the world (Nowicka and Rovisco 2009: 3). Migrants are understood to embrace and mobilize particular values and ideas to different degrees.

Narratives are a crucial element of everyday life. 'To live without narrative (…) is to live in an essentially meaningless perpetual present, devoid of form and coherence; it is to experience the world as disconnected and fragmented, an endless series of things that happen' (Freeman 1993; Gerber 2006: 68). Narratives are created by individuals in their social *milieu* and are determined by the dominating discourse. 'The extent to which autobiographical memories are stored as narratives is an open question (…), but whether told to oneself or to another, autobiographical memories are usually told. Thus, the structure of discourse affects the structure of recall, which in turn affects the structure of later recall' (Rubin 1995: 2). The discourse, in turn, is both shaped by the collective memory, as well as contributes to the changes of the collective memory. 'It is in society that people normally acquire their memories. It is also in society that they recall, recognize and localize their memories' (Halbwachs 1992: 38). To put it simply, personal recollections depend on individual experience, but the way they are rationalized, phrased and memorized is embedded in a cultural *milieu* in which the individual lives.

In migration and diaspora studies, narratives are an important source to examine insights of everyday life within a diasporic people. The analysis presented in this chapter aims to deconstruct migrants' narratives and to demonstrate the ways in which migrants' memories were shaped by the dominating discourses and the collective memory at the time of their migration and after. It is intriguing to trace the way in which these authors ascribed meaning to events in both host and home societies. No doubt, '(…) meanings are not always clear when an event occurs but often must be categorized and elaborated later' (Rubin 1995: 9).

This chapter does not claim to be representative for every Pole living or having lived in Berlin, but rather presents a detailed analysis of the memoirs and narratives of representatives from both groups (old and new) of Polish migrants in Berlin. Two memoirs were selected as samples for the 'old' migrant narratives.[1] The first one was written by Marta Czebatul (1999), a Polish working-class woman. To date, her account has not attracted scholarly attention. In contrast, the second memoir by Władysław Berkan (1924), who was one of the leaders of the local Polish diaspora in Berlin, has attracted substantial attention and is often cited by Polish historians. As for the narratives of recent Polish migrants, a variety of sources are used,[2] such as published interviews with Poles who live in Berlin, their written statements, recollections and interviews conducted as a pilot study.[3]

Poland and migration

Poland is a country populated by more than 38 million inhabitants, and its diaspora is estimated at about 15 million. A mass outflow of migrants started in the mid-nineteenth century, reaching its peak at the beginning of the twentieth century without stopping since then (Praszałowicz 2011). Although many Poles migrated to the United States, right from the start of Polish migration, Germany was the most popular European destination for Poles. Within Germany, Berlin and the Ruhr Basin became the largest centres of the Polish diaspora, although both areas experienced an outflow of many Poles in the inter-war period. Noteworthy here is the fact that those Poles who decided to stay experienced suppression during the Second World War (Stefanski 2009).

Post-war migration waves increased in the 1970s and accelerated in the 1980s and 1990s. These population movements included both *Aussiedler* and *Spätaussiedler* (individuals who entered Germany as German natives, and automatically acquired German citizenship) as well as other migrants from Poland, documented and undocumented. In particular, the *Aussiedler* have struggled with their identity, reinventing it. In this chapter, there is no room to elaborate on this complicated process of renegotiating their identity. Yet, it is important that this group is taken into account, because many *Aussiedler* keep in touch with the 'old country', and retain at least some elements of their (former) local or regional identity.

Polish Berlin

At the beginning of the twentieth century, the Polish population in Berlin was approximately 100,000. Until 1918, there was no state border separating Brandenburg with Berlin from Prussian Poland. Therefore, although literally international (from Polish territory to German territory), the Polish migrant influx to Berlin was not an inter-state move at the beginning of the twentieth century. In the inter-war period, the number of Poles in Berlin declined to just 30,000. Today, again, it is estimated that approximately 100,000 Poles live in Berlin (Praszałowicz 2010).[4] In comparison to Polish immigrant communities in other European cities, like Paris or London, Berlin became a unique centre of the Polish diaspora, mainly because of its close proximity to Polish regions such as Poznania, Silesia and Pomerania. To date, for Poles, Berlin is the closest Western metropolis which offers good job opportunities in both sectors of its dual market economy (Piore 1980). Most recently, with Poland's accession to the European

Union (EU) in May 2004 and with the opening of the German labour market for Poles in May 2011, Poles can easily settle and work in Germany.

Noteworthy here is that despite significant differences between Polish and German culture, in the past there was no substantial cultural distance between Polish migrants and the city's residents. Polish migrants, most of whom were Prussian subjects until 1918, usually spoke German and were exposed to German values and customs. In other words, before entering the German *milieu*, they had acquired cultural capital which helped them to find their way into the host society. Notwithstanding, the sense of alienation prevailed in some memoirs of past Polish residents of Berlin, and can be traced in the narratives of recent migrants.

The Polish immigrant community in Berlin has become diverse by both migration inflows in the past as well as present. In the past, it represented all regions of Poland, and all social strata. It included the aristocracy, intellectuals, Polish members of the Prussian *Landtag*, and the German *Reichstag* from 1871, tradesmen, entrepreneurs, clerks, artisans and workers. Each social group was further internally diversified. Immigrants from Prussian Poland and working-class migrants prevailed amongst the migrants. Today, Polish Berlin is also diversified: its higher strata is represented by scholars and artists, businessmen and computer scientists; the middle strata by Polish shopkeepers, nurses and plumbers; and the working strata includes employees in the construction sector, house renovation and domestic services. Yet, labour migrants in the lower strata dominate among these Polish migrants.

In contrast to Polish diaspora centres in other Western cities (Walaszek 2001), Berlin Poles have not formed one or more ethnic neighbourhoods, but have settled throughout the city with a few concentrations in eastern Berlin. Throughout the history of the Polish diaspora in Berlin, the immigrant population has shared public spaces with the local German and non-German population. This is not to say that there was no internal group activity. Poles have established a variety of Polish voluntary associations in Berlin, and this ethnic cohesion has been instrumental in sustaining and negotiating Polish national identity, fuelled in the past by the national Polish–German tension.[5] Despite the large number of Polish associations and national activities it must be noted that the majority of Poles did and do not participate in migrant community life. At the beginning of the twentieth century, Kazimierz Rakowski, one of the local Polish community leaders, estimated that only 5,000 Poles joined ethnic associations in Berlin or read Polish newspapers (Rakowski 1901). The contepo-rary situation is probably the same; despite numerous Polish societies, most Polish migrants do not become active members of these ethnic associations. One of the reasons could be the geographical proximity to Poland, which results in regular contacts with folks at home and so there is perhaps no need for such active participation. Another explanation can be found in the often temporary character of past and present migration that makes Polish migrants indifferent to the ethnic cause.

Memoirs – 'old' migration narratives

As outlined in the introduction to this chapter, two memoirs are examined in the context of the narratives of 'old' migration processes. The first memoirs were written by Marta Czebatul (1999) who was born into a Polish working-class family in Germany and lived for several years in Berlin. The most interesting aspect of her migrant narrative is that she moved from Berlin to her parent's home village in Poznania[6] and back to Berlin several times in her life. The second memoirs were written by Władysław Berkan (1942) who originated from Pomerania, an eastern Prussian province, lived in Berlin and successfully advanced to the metropolitan middle class in Berlin, but returned to Poland after the First World War.

Marta Czebatul

Marta Czebatul was born in 1907 in a Polish family, in Zehdnick, Brandenburg. Her father Maciej Szułcik, an unskilled or semi-skilled worker was employed there in a brick factory. He arrived in Zehdnick from the province of Poznania with his wife and baby son one year after he got married. In her memoirs, Marta does not provide the exact year of their arrival. It seems that Marta's father had worked in Zehdnick before he got married. His wife Agnieszka (nee Woźna) came from his home village of Bolewice in Poznania. For both Marta's parents Polish was their mother tongue, but they also spoke German – a language they learnt at school in Poznania. They had two more children while living in Zehdnick: a son born in 1909 and a daughter in 1910. Just before the start of the First World War, the family moved from Zehdnick to Berlin. Marta's father served in the German army during the war, and died on his way home when the war was over. Her mother found employment and stayed in Berlin until the 1930s when she retired and settled in a small Prussian town near the German/Polish border. Today, this town is called Świebodzin and is located on Polish territory. During the war her younger son moved to a small town Wołejkowicze near Vilnius which was then part of Polish Lithuania, while the other two children stayed in Berlin. The oldest son became a clerk in the Polish consulate, and the youngest daughter married a Pole who was employed in a Berlin branch of a Polish travel agency called *Orbis*. Marta, who was less educated than her sibling and had fewer chances to progress in the job market, thus, commuted between Berlin, Poznania and Wołejkowicze for a few years.

Despite living for many years in Germany, Marta's parents lived transnational lives. They managed to keep in touch with their relatives in Poznania and also thought of Poznania as their *Heimat* (homeland). Marta spent only a few years of her childhood in Poznania, but came back to Poznania as a teenager where she stayed with her aunt and uncle. The couple had no children and apparently helped Marta's mother when she became head of the household following the death of her father. Marta did not provide any details of the arrangement in her memoirs, except to say that while she attended an elementary school in Berlin (1916–1919) her sister was staying in Poznania. After Marta completed elementary school she was sent to Poznania and stayed there for a few years until 1924, while her sister was taken to Berlin. Marta spent four more years in Berlin from 1924 to 1928, where she attended a vocational school and worked as a salesperson. She left Berlin for good in 1928. In 1932, she married a Polish farmer who was her brother's neighbour in Polish Lithuania, and moved to Wołejkowicze. The couple had two children before the Second World War began. By the spring of 1944, Marta sought shelter in Germany, when her house was destroyed (burned down), and her husband was detained in a Nazi prisoner of war (POW) camp. She was employed on a farm, where she stayed together with her children. The farm owner (a woman) even managed to arrange for the release of Marta's husband from the POW camp, so that he could rejoin them and work on the farm. After the war, Marta and her husband decided to return to Poland and settled initially in the town of Trzciel, and later in Międzyrzecz, located on former German territory.

This brief outline of Marta's biography shows how much she moved around in her life between the regions of Brandenburg, Poznania and Polish Lithuania. Poznania is a German/Polish borderland, and Vilnius with its vicinity is a Polish/Lithuanian borderland. The population in these territories was ethnically mixed and national tensions were self-evident in the first half of the twentieth century. This 'ethnic' situation shaped the dominant discourse in which ethnic/national belonging was emphasized and gave it a highly nationalistic tone. Moreover, the communist propaganda of the post-war decades skillfully used post-war anti-German feeling in Poland in order to strengthen the authorities. Marta wrote her memoirs at an elderly age, and while narrating her youth, she recalled it through the prism of her later experiences of this type of nationalist discourse. Therefore, her narrative gives a taste of the Polish nationalistic discourse ('the ethnic absolutism').

According to her own testimony, Marta did not face any major trouble or disappointments in view of her Polishness during her years in Berlin. In fact, there were several instances which documented how easily it was for Marta to become integrated into German society. For example, in her memoirs Marta mentioned that during the First World War, her aunt arrived in Berlin from Poznania to provide food from her farm. The food not only served Marta's family but also their German neighbours, who were eager to buy it. This situation uplifted the status of Marta's Polish family amongst their German neighbourhood in Berlin. Another example of her easy integration into German society can be found when she returned to Berlin from Poznania and was enrolled into a vocational school. She wrote how worried she was about her German language skills, which had deteriorated during her stay in Poznania. However, on her first school day, a teacher asked her to sit in the classroom next to a Polish girl. A Polish classmate with fluent German was instructed to help Marta. Moreover, she befriended a number of German girls and invited one of them to attend a service in the Roman Catholic church. After many decades, she recalled this event with nostalgia, appreciating the fact that her Protestant friend went with her to the

Catholic Church. When Marta dropped out of this school for an unknown reason, she found a job in the German neighbourhood as a shop assistant. She liked the job, and did not recall any problems with her employer.

In her memoirs, Marta recalls that many Poles in Berlin assimilated, but that she felt upset about it. Her family regularly attended St. Sebastian Church in Berlin where Polish sermons were held every second Sunday. They met and befriended some Polish workers in Berlin and noted that they were not eager to teach their children their native language. Marta's mother was especially upset that a certain Jagiełło couple, whom she had visited on her way back from the church, spoke only German to their four children. The special reason for disappointment was the family's name: 'how can they be Jagiełło, if all their children are Germanized?'(Czebatul 1999: 27). Jagiełło was the name of a Polish king (1386–1434), the first of the Jagiellonian Dynasty, celebrated in the national tradition, and clearly perceived by Marta's mother as an embodiment of Polishness. Indeed, Jagiełło led the royal army of the Polish Lithuanian Commonwealth to victory over the Teutonic Knights at the *Tannenberg Schlacht* (battle) in 1410. The victory (ignored by German historians) became one of the key notions of Polish collective memory and is often used in nationalistic anti-German discourse. Noteworthy here is that the king Władysław Jagiełło was actually ethnically Lithuanian. Clearly, Marta's mother did not think about Jagiełło's origin or the complicated roots of royal dynasties.

Although Marta was well integrated in Berlin, at times she declared that she hated Berlin and Berliners, even though in her memoirs there is no evidence of any negative experiences during the years she spent in the city. She probably believed that a good Pole should hate Germans because of Poland's experiences during First and Second World Wars and, hence, try to follow this national Polish discourse. Despite the prospects of possible Soviet oppression, she did not hesitate to seek work in Germany at the end of the Second World War. After the war, she insisted on returning to Poland. She did not return to her home village though, but went to another part of Poland, which belonged to Germany before the war. She complained that she arrived there too late to profit from the affluent homesteads left by the local German population when they were forced to resettle in Germany after the Second World War. At the end of her story, Marta added that in the 1980s, her grandson moved to Germany as an *Aussiedler*.

There is no doubt that Marta practiced some kind of cosmopolitanism in her everyday life in Berlin by engaging with the other, yet never losing sight of her Polishness. In her memoirs, she stressed that her family emphasized their 'Polishness' in Berlin. Key elements of her 'Polishness' that she emphasized in her memoirs were, amongst others, Polish language maintenance, traditional Polish Catholicism, Polish immigrant community building in Berlin and keeping in touch with her relatives in Poznania. Indeed, close contacts with their relatives back home helped Marta's family to retain the sense of belonging to the home community. She praised her parents' patriotic attitudes and the way her father was involved in a migrant association. Yet, at other times, it seems that her interaction with the German 'other' at school, work and in the neighbourhood was closer and more intimate than she was ready to admit. Her migration narrative demonstrates that her cultural identity was not 'fixed' but rather entwined within a complex web of cultural attachments. Following Beck (2006), one could say that her everyday life was not composed of either Polish or German culture, but of both Polish and German culture.

Władysław Berkan

Władysław Berkan was born in 1859 in Pomerania (village of Sampława) as the son of a shoemaker. In his youth, he was an apprentice in several tailor workshops in two small local towns – Lubawa and Grodzisk – in which Poles, Germans and Jews worked together. He did not learn German at school, but took a course later on to master the language. In 1880, he migrated to Berlin with a German friend from his home village. He worked there as a tailor. In the late 1880s, he established his own workshop, married a Polish migrant woman, and invited his younger brother to join him in Berlin. According to his memoirs, the workshop became a successful business: the turnout doubled each year. Berkan stayed in Berlin until 1917. In 1917, he passed on his business to his brother and moved to Poznań. Because of the inflation and the currency exchange, he lost all his savings.

Berkan's memoirs are an interesting source for information on the lives of Polish immigrants in Berlin. During his Berlin years, Berkan was actively involved in local migrant organizations and became a well-known Polish leader. In his memoirs, he created himself as a person who dedicated his life to the Polish cause. Berkan did not hesitate to announce highly judgemental opinions on the local Polish diaspora, its endeavours, political activities, sport and even private immigrant life. In contrast to Marta, he did not focus as much on his everyday life in Berlin, but rather emphasized Polish cultural continuity in Berlin and presented his opinion on local Polish immigrant societies. As was the case with Marta's memoirs, Berkan often used elements which played a key role in the Polish nationalistic narrative: Polishness (cultural continuity), language maintenance, traditional Polish Catholicism and Polish immigrant community building in Berlin. In contrast to Marta, he did not elaborate on his parents, who stayed behind in their home village. However, he did stay in touch with them, and as already noted, he brought his brother to Berlin and made him his business partner.

Berkan did not provide much information about his private life. He emphasized, however, that he was determined to marry a Polish woman and that he was looking for a proper candidate within the local Polish diaspora. He eagerly elaborated on the importance of marrying within the Polish group and on the necessity of passing on one's mother tongue to the younger generation. He agreed with his wife about speaking only Polish at home, and they both encouraged Polish language education for their children. In his memoirs, he used his family as an example to be followed. Berkan criticized migrants who married outside the

Polish community, even if Polish males were successful in 'Polonizing' their German wives. Clearly 'Polonization' was right, and Germanization was wrong from the author's point of view: 'Why to marry German women and then to bother [trying to Polonize them – DP], if there are so many of our Polish women around. I do respect any nationality; nevertheless, I prefer my own' (Berkan 1924: 180).

In his memoirs, he also commented on changes in Polish society as well as within the Polish diaspora. For example, he opposed higher education for women: 'Nowadays many women in Poland study. I suppose they do because they have no chance to get married....' (Berkan 1924: 262). As a conservative activist, he fought against the working-class movement and promoted Polish Catholic organizations. Although he was a devoted Pole, he stressed that he was always a loyal German citizen. Indeed, he was a spokesman for the Polish immigrant population, and a broker who mediated between Polish communities and both German and non-German communities in Berlin. There is no doubt that Berkan appeared to be a cosmopolitan patriot (Appiah 1997), practicing cosmopolitanism through national lenses. His cosmopolitanism involved close contact with both Germans and foreigners in Berlin. Aside from Germans and Poles, there were many Scandinavians among his customers. Moreover, he recalled cooperation between Polish and Czech societies in Berlin and represented the Berlin-based Polish Industrial Society at the opening ceremony of the National Theatre in Prague in 1886.

Berkan's memoirs are far more representative of the Polish diaspora than Marta's. Marta's memoirs give an unusual insight into the lower-class migrant population in Berlin, provided by a woman, and therefore representing a forgotten perspective. Berkan, as an activist in Polish life in Berlin, felt it was his duty to write generally about the Polish diaspora in Berlin. Both in his life and in his memoirs, he presented himself in a self-conscious and reflective way. While writing about Polish assimilation in Berlin, he felt, he knew the facts, and believed he was in a position to assess them. He openly rejected assimilation, assuming that it entailed uprooting.

Marta's and Berkan's stories demonstrate that the 'old' migration memories were embedded in the Polish national (anti-German) discourse. The authors focused on Polish ethnic community-building and cultural continuity. The authors perceived Polish 'ethnic' activities as a weapon against the assimilation process. Yet, both migrant memoirs show that their everyday experiences were defined by an interaction with, and openness toward, other cultures, in particular German culture. In this way, their memoirs seem to confirm what previous research on cosmopolitanism has stressed: that cosmopolitan people can combine strong ethnic affiliations with an attitude that recognizes cultural plurality, making it easy to move around different cultural systems (Amin 2004; Appiah 1997). While in some ways, Berkan's and Marta's identities indicate an essentialized sense of ethnic belonging, which emphasizes Polishness, in other ways they exhibit a complex interweaving of sameness and otherness.

New migrations, new narratives

Nowadays, the old Polish narratives still appeal to many Poles, including Polish migrants. At the same time, new notions of Polishness pertain to the public discourse. They reflect recent political and social developments, such as Polish accession to the EU and the associated opening of the labour market and consequent Polish migration flows. The new migration discourse generates new ways in which migrants try to grasp and define their experience. By the late 1990s, the number of Poles in Berlin again reached 100,000. The number, based on estimations, embraced both regular and irregular Polish migrants. As was the case in the past, Polish migration constitutes a variety of streams which produces a number of communities, while immigrants still cluster in certain (lower) sectors of the segmented economy. However, it is noteworthy that these diverse Polish communities interact with each other too. When asked to narrate their everyday experiences in Berlin, in most cases, Polish migrants explicitly reject the old national discourse. Their stories are embedded in the discourse in which attention is paid to the social, political and cultural changes as well as to cultural pluralism and mobility. One the one hand, migrants perceive Berlin as a metropolis of global culture; on the other hand, as a mosaic of cultural districts, where each district gives its inhabitants a sense of belonging and locality (Kerski 2008).

While Berkan and Marta used Poland and Germany as their frame of reference for their memoirs, today's Polish migrants' refer much more to the local level, the city itself. Such a localism started well before the end of the cold war. Under communism, West Berlin was defined by some Polish newcomers as '*Eine Insel der Freiheit*' (an island of freedom) (Krynicki 2008–2009), addressing major political discourses in their narratives. Indeed, today many Polish migrants declare they have settled down in Berlin because the city makes them feel free and provides an opportunity for self-fulfilment. Some Polish migrants define 'their' Berlin in diasporic terms, and assert that in the 1980s, Berlin became a true 'Polish city' (Szaruga 2008–2009). In this respect, once again, the political context of Polishness is stressed. The well-known writer and poet Leszek Szaruga writes about the Berlin experience of Polish refugees and their anti-communist activity, which was possible during the cold war. After 1989, Poles eagerly elaborated on their everyday life in Berlin, focusing on their individual impressions: for example, praising the city's public spaces in which everybody feels welcome (Tokarczuk 2008–2009). During her pilot survey, conducted in 2008, the author asked a Polish artist why he chose to settle in Berlin. The artist declared in a provocative tone – 'Berlin is the only city in Germany'. A journalist who originates from Warsaw remembers that when she visited Berlin for the first time, she decided she wanted to move there (Praszałowicz 2010). Many migrants indulge in the alternative culture, which has an international, multicultural imprint, and for which the city is famous.

Many scholars describe contemporary migration flows as transnational movements. Indeed, most temporary Polish migrants live in two or more social and cultural spaces at the same time, and they feel attached to two or more homes. They reinvent and reconstruct their identity, making conscious choices and living both in their home and foreign culture. Contrary to the 'old' discourse, the 'new' Polish migrant narrative is a product of a globalizing world in which multiple national identities and transnationalism are perceived as natural. Polish migrants do not as much elaborate on ethnic activity, even if they are involved in it: rather, during our interviews, they talk about changing their transnational sense of belonging. Representatives of recent Polish migration no longer perceive only two options for their identity, Polish or German: rather, during our interviews, Polish migrants declare a more complex identity – 'I am a Jew from Poland', or 'I am German and Jewish and Silesian' (Praszałowicz 2010). The latter declaration came from a woman who is *Aussiedlerin.* In an interview, another Polish woman talks about her European identity and refuses to identify in national terms. Noteworthy here is that this woman's husband originates from a non-European country, which might reinforce her European sense of belonging. On the other hand, it must be noted that the couple feels more comfortable in Berlin than in Poland, where her husband is likely to encounter more prejudice.[7] Another migrant from Poland who has multiple ancestry (Polish and non-European) elaborates explicitly on his reinvention of identity, which was made possible when he settled in Berlin:

> The ethnically and socially diverse, yet co-existing elements helped me as a Polish-Iraqi migrant from Gdansk to settle in this metropolitan city. I didn't feel like I have to live in a ghetto. On the contrary, those Germans, who have Polish, French or Flemish names and live in this metropolitan city, see me as a proper *Berliner* – a Slavic immigrant (*Einwanderer*) and child of an oriental guest worker (*Gastarbeiter*). Only in this city, I learnt that I don't have to choose between my Polish and Iraqi roots and no one demands that I have to pretend to be a proper German. Such a demand would only make me feel like a stranger and create a feeling of inferiority. My complex identity is part of this city's history and is one of the few elements that have stayed the same in Berlin.
>
> (Kerski 2008: 421)

Scholars point to the fact that within the Polish diaspora in Berlin one can discern a whole variety of identities, ranging from an essentialized Polish identity to a true denial of Polish identity (Morawska 2008). Moreover, migrants who participate in the life of the local Polish diaspora in Berlin develop new forms of Polish community life. For example, they have established a Polish Rotary Club in Berlin and, on the opposite side, Club *Polnischer Versager* (Polish Losers' Club). The latter one operates an alternative theatre which is popular far beyond the diaspora boundaries. In fact, both the club and the theatre are truly cosmopolitan. Their members reach for provocation and grotesque Polishness; they distance themselves from the old forms of Polish national expression and often ridicule traditional Polish discourses. There are also Polish squatters in

Berlin who did not establish any society, but developed a strong sense of group belonging. The identity and narrative of these groups are far from the old forms and have nothing to do with 'ethnic absolutism'. At the same time, the traditional structures of diaspora, including the Polish parish, press, cultural and professional clubs, still play an important role. Some of their members declare an identity that can be named 'ethnicity forever' (Morawska 2008). There are also migrants who assert that, nowadays, Berlin is just a place to live, and they point to the similarities between living in Poland and in Germany (Niewrzęda 2005).

Conclusion

For many years, migration scholars took 'old' migrants' memories literary and believed that the main concern of Polish migrants was sustaining their cultural tradition. Integration processes and possible consequent cosmopolitan practices and attitudes were ignored, despite the fact that most migrants displayed them. Moreover, it was assumed both by scholars and by diasporic leaders that Polish migrants' identities were bimodal, either Polish or German. This clear-cut division was constructed in Polish national discourse at a time of rising nationalism in European countries. For centuries, the nature of the Polish diasporic and even home community has been uncertain because of several political events, including the three partitions of Polish lands, two world wars, and communist rule with its aggressive anti-German propaganda. Although Poland was independent between 1918 and 1939, as a country it only regained its complete independence in 1989. Thus, one could assume that this long-lasting feeling of insecurity compelled Poles to declare their devotion to the national cause.

As a consequence of these political developments, and as Berkan's memoirs have demonstrated, Polish migrants perceived their task as to raise the issue of Polish independence on the international political stage and to convey and explain Polish aspirations to a wider audience in the host society. This is how the national mission was understood by the leaders, both on Polish lands and within communities of Polish diaspora, for many years. Historians and devoted members of the Polish intelligentsia felt obliged to support the cause and to find means to compensate the feeling of insecurity. Therefore, studies on Polish migrations and the diaspora were to give a hope that Polish national identity would be carried on by Poles in exile (Praszałowicz 2010).

From 1989 onwards, with the opening of borders and greater mobility, Polish migrants have distanced themselves from nationalistic discourse. In this new atmosphere after the end of the cold war, it has become easier for migrants to integrate, and to realize that there are some universal paths of cultural exchanges of migrant communities: hence, to positively engage with 'the otherness of the other' and the oneness of the world (Nowicka and Rovisco 2009: 3). In other words, displaying cosmopolitanism attitudes openly alongside living out Polishness has become part of everyday life. Today, Polish migrants are able to embody several identities. They can identify as Jews from Poland or

Silesian/Jewish. Their narratives cease to be Polish-centred; instead, 'new' Polish migrants occasionally draw on cosmopolitanism as an identity resource, which denotes a stance toward diversity that enables them to construct belonging in terms of ethnicity, as well as multicultural location.

Notes

1 The primary sources used to explore 'old' migrant narratives include Abramowicz (1979), Berkan (1924), Czebatul (1999) and Rose (1932).
2 The primary sources used to examine recent narratives of Polish migration include Danielewicz-Kerski and Górny (eds) (2008), Gotfryd (2005), Kerski (2008), Krynicki (2008–2009), Niewrzęda (2005), Szaruga (2008–2009) and Tokarczuk (2008–2009).
3 The author conducted 15 interviews in October 2008. The study was supported by the Polish Ministry of Higher Education (2007–2009).
4 For more detail on the notion of 'Polish Berlin' see, for example, Praszałowicz (2006, 2010).
5 For example, in 1919, there were 56 Polish voluntary associations in Berlin (Praszałowicz 2006).
6 Poznania is a Polish region within the borders of the former Prussian Poland.
7 For a discussion on attitudes towards foreigners in Poland, see studies by Nowicka and Łodziński (1990) and Nowicka (2011).

Part IV

Exploring ethical challenges in research on migration

7 The beginning and end of a beautiful friendship

Ethical issues in the ethnographic study of sociality amongst Russian-speaking migrants in London

Darya Malyutina

Considering this chapter, I am delighted with an opportunity to write about the drawbacks of doing qualitative research with a diverse set of people, trying (or having) to become a part of their lives and listening to what turned out to be emotional accounts of their lives. I remembered one of dozens of fieldwork cases involving an emotional interaction between myself and one respondent. As soon as I started to describe the situation, I realized that I had not mentioned it in the research diary. Something was preventing me from elaborating the topic: Too personal, either for me or for the other participant in the situation? What if she reads my chapter? And, more importantly, does this case have anything to do with my study of friendship, or with the pitfalls of an actual relationship? Can the emotions produced by one's friendship with respondents be counted as part of the research outcomes, or do they relate to me as a person only?

This chapter is a part of a study of informal social relationships amongst Russian-speaking migrants in London. Being a Russian in London myself, I researched friendship and the formation of personal networks in this migrant group, in order to understand how informal connections can trigger the development of this community in a global city. By analysing the complexity and inner diversity of this migrant group, I want to demonstrate the limitations of developing a tightly knit community, and critically analyse the application of the concept of transnationalism to contemporary migration. My work also investigates relationships with non-Russian speakers, studying the dynamics of migrants' cosmopolitan and racialized attitudes. It shows how sociality works in different social groups of migrants. I chose friendship as my key concept, as the basis of personal network formation, although this does not have to be limited by kinship, neighbourhood or formal occupation, but seems to be conditioned by national/ethnic origin amongst former Soviet Union (FSU) migrants. In order to study sociality in a stratified population of a global city, I use ethnographic methods, which means I spend a lot of time in close communication with my informants.

Ethnography is connected with social relationships unfolding in the context of the exchange of information, and the personal involvement of the researcher (Agar 1980). 'The researcher cannot conveniently tuck away the personal behind the professional, because fieldwork *is* personal' (England 1994: 249). I understand 'ethically important moments' as 'difficult, often subtle, and usually unpredictable situations that arise in the practice of doing research' (Guillemin and Gillam 2004: 262). This chapter aims to analyse the ethical drawbacks of qualitative study done by a researcher with an implied 'insider' position, drawing on the example of Russian-speaking migrants in London. It presents an analytical representation of relationships with respondents and a personal account derived from the fieldwork. First, I approach the issues of establishing relationships with respondents, concentrating on the researcher's positionality. Secondly, I present a brief analysis of friendship amongst Russian-speaking migrants as a means of forming a particular community, and reflect on its methodological implications for the researcher–researched relationship. Finally, I focus on the problems posed by particular relationships, and conclude with discussing the limitations of doing qualitative research with a presumably 'insider' status. There are three key points of this chapter. First, ethnographic research on migration, especially when done by a migrant researcher, can benefit from feminist traditions, particularly on the issues of reflexivity, positionality and power inequalities. Secondly, I support the feminist argument of making the downsides of the research relationships clear, and argue that building a relationship with respondents is in itself a way of practically studying a community. Finally, I claim that belonging to the same national/ethnic community can be helpful for the research; however, being close to respondents does not make one an insider, as a whole range of differences exist between the researcher and respondents.

Reflexivity and positionality in qualitative research: the feminist approach

Following feminist geography, this chapter attempts to 'try and make more visible the mystery that is the research process, without "drawing a veil over the implications of [the researcher's] own position"' (Rose 1997: 309; McDowell 1992: 403). If we are studying friendship by means of ethnography, and want to develop reflexivity, we should make visible the relationships between the researcher and the researched and the implications of these for the relationships in the migrant community. The main body of literature I rely on includes writings on qualitative research and ethnography (Agar 1980; Clifford and Marcus 1986; Denzin and Lincoln 2005); migration and transnationalism literature (Conradson and Latham 2005a, 2005b; Glick Schiller *et al.* 2011; Vertovec 2007); works on Russian migrants and Russian friendship (Byford 2009; Kharkhordin 1999, 2009; Shlapentokh 1989); and feminist geography (England 1994; McDowell 1992; Rose 1997).

Participant observation, as opposed to interviewing, requires continuous involvement of the researcher in the life of a community, and often implies developing friendships. In my case, it was even more complicated because I often personally experienced friendship while also researching it. While maintaining close friendly relations with the informant can give access to deep layers of personal information, specialized knowledge and richer communication, it can bias the ethnographer's interpretations, as a researcher's role is gradually replaced with that of a friend. Judging respondents from the point of view of a friend may influence the accuracy and validity of research. Keith (1992: 553) mentions that moral, political and ethical considerations that structure the context of research 'involve instant judgment in the field and rationalization of such judgment in report: [...] the representational accuracy of the report is structured by the moral (politico-ethical) assessment of particular situations'. Hence the double risk: judgement in the field may bias your interpretation; personalized interpretations may bias the academic representation of your work.

The issue of power inequalities often arises in feminist research (England 1994; McDowell 1992; Rose 1997). The researcher's position is connected with the production of knowledge about others, and the risks involved imply that knowledge obtained via this kind of power cannot be considered universal. A way of reducing overgeneralizations is 'making one's position known, which involves making it visible and making the specificity of its perspective clear' (Rose 1997: 308), which involves a reflexive approach both 'inward' towards the self of the researcher, and 'outward' to her research and the world. Reflexivity can help to avoid essentializing the distinctive qualities of the researcher, and thinking critically about the research process (Guillemin and Gillam 2004; Robertson 2002). However, agnostically admitting that those researched can never be fully understood, as they are different from the researcher, Rose (1997: 317) underlines that the research process is a 'fragmented space, webbed across gaps in understandings, saturated with power, but also, paradoxically, with uncertainty'. Reflexivity is considered a tool for making the asymmetries visible, but not for removing them (England 1994: 250).

The insider–outsider question has been a problematic topic in social sciences (Agar 1980; Merton 1972), and is particularly relevant to migration research and feminist geography (Abu-Lughod 1991; Ganga and Scott 2006; McDowell 1992; Mullings 1999). Defining the researcher's position as an insider or an outsider is difficult, and the relative benefits of either definition are disputable. Sharing the same nationality with the respondent is in most cases not enough to be considered an 'insider', as social status includes other criteria that may make people reject an insider status – class, gender, age, legal status, marital status, etc. Also, the 'insider/outsider' binary 'is not only highly unstable but also one that ignores the dynamism of positionalities in time and through space' (Mullings 1999: 340). Considering the historically produced and contested nature of cultures (Clifford and Marcus 1986: 18), and cultural identity being 'not an essence but positioning' (Hall 1990: 226), defining one's own niche in a certain community can be far from straightforward.

This discussion points to the idea that qualitative research implies the researcher's personal involvement in the lives of her informants. Extending the researcher's social position outside the boundaries of a professional role brings to the fore not only the issue of power inequalities between them but also the transience of positions. Hence, the problems of identifying one's position as insider or outsider: the multiplicity of socio-economic, cultural, generational, legal, gender and other differences require reflexivity to be employed throughout the research in order to constantly analyse and redefine one's positionality. Although attempts to be reflexive cannot possibly make us understand the respondent completely, they are helpful in pointing out the differences between the participants of the interaction. This discussion leads to an idea that a feminist approach to research can be helpful for qualitative migration studies, particularly in methodological and ethical questions.

Approaching Russian-speakers in London

The characteristic of London that makes it popular for migration studies is manifested in its highly stratified population, constant influx of migrants, anonymity of social life, and presence of many cultures and languages. 'Super-diversity', as Vertovec (2007: 1024) puts it while describing the contemporary state of London, is 'a dynamic interplay of variables among an increased number of new, small and scattered, multiple-origin, transnationally connected, socio-economically differentiated and legally stratified immigrants who have arrived over the last decade'. Miller (2008) names London as 'nowhere in particular', a place where people come 'as an escape from [...] identity by origin rather than to replace one with another'. New immigrant groups have appeared on the migrant scene of Britain, differing from the previous migrants from Commonwealth countries or former colonial territories by being smaller, less organized, and highly differentiated, and not having similar historical links with Britain (Vertovec 2007: 1029). Migration from the FSU countries in its current volume, regularity and variety represents one of the recent trends that contribute to diversification of social life in London.

Russian-speaking migration has not been the focus of much general migration research, although it is achieving popularity as a topic of interest for Russian and East European Studies (Byford 2009). In recent years, an increasing number of postgraduate students have embarked on studying this group from different analytical angles, themselves being part of the current migration wave from Russia and other FSU countries. However, migration literature looking at the UK does not pay much attention to Russian speakers. This is partly due to the relatively recent appearance of this migrant group in the UK in noticeable numbers, which are still unclear in the statistics. In addition, most of the FSU countries are not member states of the European Union, apart from Estonia, Latvia and Lithuania – countries which are beginning to receive mention in the migration literature (McDowell 2008). In addition to the tendency of academic work on Russian speakers to group together, rather than proliferate into the general sphere of migrations studies, it causes the relative academic 'invisibility' of Russian-speaking migrants. This 'invisibility' has a more mundane dimension, which is represented by them being physically undistinguishable from an average 'white European', and not forming a diaspora in a classic sense (Cohen 1997). Kopnina (2005) underlines that Russian speakers tend to form 'subcommunities' on the basis of social class, profession or interests, and to engage in international networks. She explains this to be the consequence of a low critical mass and the lack of physical difference between migrants and other white inhabitants of the UK. Internal divisions 'account for the fragmentation of Russian migrants and their lack of interest in or awareness of the "Russian community"' (Kopnina 2005: 13). Byford (2009: 55) also argues that 'diasporization' of the post-Soviet Russian migrants in Britain is not based directly on Russian ethnos, state, national culture, or even language, but rather united by 'a historically-specific socio-cultural background shared by the generation of people born in the former USSR [...], whose formative identifications are therefore rooted, somewhat peculiarly, in a state and society that are no more, and whose life-worlds span the distinctive juncture between late socialism and postsocialism'. This migrant group is characterized by fluidity of boundaries, implying its relative openness to representatives and institutions not of post-Soviet origin; high degree of social stratification; and different extents to which migrants rely on their community (Byford 2009). Also, the Russian-speaking population is geographically scattered around London, settling according to their social class rather than their proximity to compatriots.

Thus, this research faced the issue of studying friendship in a diverse and stratified 'community', focusing on the constitution of migrants' personal networks and relationships inside them.[1] Ethnographic study of the Russian-speaking migrant community in London included participant observation in a Russian bar and 35 semi-structured interviews with migrants selected by the snowball or referral sampling technique. Respondents with different backgrounds and social statuses were selected in order to achieve a multifaceted picture of the community.[2]

The practical implementation of the technique marked the rapid, multidirectional, sometimes problematic proliferation of the sample and of the researcher's personal social network. The study started with participant observation in a Russian bar – one of the principal initial sources of respondents. Another source was existing Russian-speaking acquaintances. Some of the informants were found through professional contacts. Finally, a few were randomly 'picked up' in non-research-related circumstances, when I needed representatives of social groups different from those present in the sample.

The beginning

The fieldwork in many cases led to informal relationships with Russian-speaking migrants: hanging out with them, visiting each other, eating out, drinking and chatting. My ethnographic observations of the others' interactions and personal experience of becoming a part of a certain community showed that, initially, a person's status of Russian-speaking migrant promotes the establishment of intersubjectivity. Sociality developed on the basis of a common cultural background and language; those who do not possess these characteristics can be discursively and practically distinguished from the community. This was more evident in a bar study, where most of the respondents were lower-status migrants, and less true for more educated, legal migrants with higher incomes and stable occupations.

Usually, the researcher's status of a Russian in London helped – at least, in the first stages of establishing relationships with respondents. Conversations with migrants at the start were connected with finding commonalities, which promoted the development of trust, and eventually the informant giving more personal and confidential information.

> R: …Are you from Moscow yourself?
> I: No, I am from Novosibirsk.
> R: From Novosibirsk? *Ooh, I love you, my dear….* People from Moscow are like Londoners. If they cook, they cook, say, only three portions. If I cook, I make a huge pot. You can eat two–three portions, and something will be left. My friend Natalia from Moscow often invited us. She was like – I've cooked four portions. I say – Why wouldn't you make more? – What would I do with it? – Well, put it in a fridge!... […] *And you are from Novosibirsk…. I've been there. One of my clients*, who is more or less *like my friend*, comes here once a year – *he's Siberian*, from Krasnoyarsk. We go to a restaurant. I've met his friends, they all are great guys. […] *My father is from Siberia.* […] And as a Siberian, he has this wide generous soul [*shirokaya dusha*], only he drinks too much…. He went to Nizhny Novgorod after his divorce with my mother. It was in 1989, you probably were not even born…
> I: I was three years old at that time.
> R: So you were born in 1986? Oh, *I went to the army in 1986…*'.
>
> (Evgeniy, 43 years, July 2009)

This excerpt, first of all, accounts for the inner diversity of the Russian-speaking migrant community in London. Evgeniy specifies people from Moscow and criticizes their stinginess, in contrast with (presumably) more generous people from Siberia. He distinguishes between clients who can approach the category of a friend, and the rest of them. A big part of his reflections on relationships with Russian speakers is lamenting about the vulgarity of his clients (Evgeniy is a property consultant for Russian millionaires) and his reluctance to engage in close relationships with them. Navigating through accounts of diversity, he manages to find similarities between himself and the researcher (italics). First of all, he parallels the researcher's origin with his experience of dealing with people from the same region of Russia; then, he remembers visiting her place of birth; next, he brings in the arguments of friendship and kinship connected to that certain place; finally, he sees a connection between two remarkable events in the lives of himself and the researcher. Although there was not much in common between us, apart from a mutual friend – his neighbour and my former colleague – similarities are constructed by the participant(s) of the interaction, facilitating the communication (for him) and the interview (for me).

Geographies of pre-migration life are often invoked in conversations between new acquaintances. Such hitherto 'invisible links' to the same city or region of origin helped a lot in getting access to respondents. Spatial closeness of places of origin usually implies the existence of specific common knowledge, common acquaintances and similarities in life experiences. My longest ongoing friendly relationship with an informant was marked by finding out about the same region of origin at the first meeting. My language teacher turned out to be flatsharing with 'a Russian guy from a city whose name I would not be able to spell, as yours – maybe you are from the same place?' who happened to be from the same city and same school with me, and ended up as a respondent. One sequence of interviews and observations started from me bumping into my former classmate in Waitrose, getting involved with his network of friends and picking up more informants. Sense of 'imagined community', or 'community of sentiment' – 'a group in which members are tied with collective sense of imagination and begin to imagine and feel things together' (Appadurai 1996: 8) were central to the fieldwork, where the links to the same places provided an additional feeling of commonality.

However, things would have been too easy if all the research connections had been simply established just because you speak the same language and come from the post-Soviet space. Russian-speaking migrants' reluctance to communicate was noticeable with some categories of people. Introversion, suspicion and refusal were also a part of the empirical research. Agar (1980: 59) regards suspicion at the initial stages of research a normal part of ethnography, as 'the ethnographer is asking for trust without yet having earned it'. In most cases it was related to social differences between the researcher and the researched. In the following example, it eliminated my chances of interviewing a potentially interesting respondent:

> I was sitting in the bar and chatting with Nadya. Suddenly, Rita came in. She was a bit drunk already, cheerfully ordered a glass of white wine and sat beside me. She was in a perky talkative mood, and immediately started a conversation. From discussing the mundane topics like weather and 'how-are-things', however, the conversation soon became more aggressive. First, she informed me of her disapproval of my eyebrow piercing and expressed a firm opinion that I should have it removed. Then, she switched her attention to my beer and told me that a girl should not drink Hoegaarden. After a while, probably after getting more drunk, she remembered what I told her when I first presented myself as a researcher. Rita embarked on accusing me of coming 'from there' and getting a place with a scholarship in a university: 'While my son could not pass the exams, could not get a place as good as yours – people like you come from your small cities in Russia and get it just like that! You are wasting the taxpayers' money, you leeches! Bugger off, go back to your Russia and do your research there! I will never give you an interview!' I was trying to stay (or at least look) indifferent during her outburst, although what I really wanted to do was pour the rest of

> her glass of wine over her head. I was calming myself down with thoughts like 'She is an aging alcoholic, with an unlucky personal life and no professional self-realization, in a deep crisis and unsatisfied, there is no point in getting sensitive for me…'. She finished her fiery speech and her wine and left the bar. The bartenders looked at me with compassion. 'Don't take it personal, that's Rita. She is always like that – saying stupid things, then feeling sorry, then getting drunk again…'.
>
> (bar ethnography excerpt, June 2009)

This situation made me think about my behaviour as a Russian-speaking bar visitor and a researcher, and the possibility of contradiction between these social roles. Her arguments related to me as a person; however, my reaction was conditioned by the requirements of my study, as I kept reminding myself that I needed that place and need not be in conflict with anyone there. Feminist theory notes that researchers are also affected by the research process, as all of the participants of interaction are involved in mutually constitutive social relations (England 1994; Rose 1997). Neglecting the researcher's emotions would be dangerous, as they inevitably influence the interpretations of social situations (Widdowfield 2000). As a person, I felt I have been treated undeservedly. As a researcher, I felt surprised – she was the first Russian I have met who was explicitly aggressive towards another Russian (who, by chance, happened to be me). It was after that situation that I realized I started to empirically prove for my sample what has been outlined in the literature (Levitt 2001; Wimmer and Glick Schiller 2003) – the Russian-speaking community, as any other, is not and does not necessarily have to be a tightly knit group of compatriots feeling affinity and solidarity towards each other, and its internal homogeneity and boundedness, and even its transnationalism, should not be overstated.

Being of the same ethnic/national origin with respondents was helpful in most cases at the initial stage of research: it promoted access and the development of trust. Further communication unfolded with a search for similarities or differences between us. Following feminist tradition, I made clear not only my success in getting in touch with informants but also my failures. I argue that all sides of the research interactions should be accounted for. Building relationships with respondents, as an implementation of research methods, in itself is a way of practically studying the community, apart from listening to narratives and observing sociality. Meanwhile, having more or less defined my place in the relationships with respondents, I went on to research and experience friendship – the kind of attitude that does not only bring but also keep people together.

'The beautiful friendship'

Researching migration in the context of contemporary global cities and their 'human dimension' (Favell 2008: 221) is increasingly linked with studying social networks, connections and relationships (Conradson and Latham 2005a, 2005b;

Scott 2006). As a way of understanding the implications and problems of lived transnationalism, cosmopolitanism and racialization in London, I focused on friendship and other kinds of informal relationships.

Migrants' communities are formed and maintained on the basis of everyday personal interactions in small social networks. Friendship for many Russian speakers is not always about connections they had before moving to London. The establishment of new social ties may be spontaneous, often with an undeniable special value assigned to such friendships. At the same time, relationships are socially stratified and selective. 'Real friendship' with compatriots often implies long-term relationships, and it is common for it to be established with those who have links with the same place of origin. However, migrants often expressed indifference or even reluctance to expand their personal networks in a transnational way while they are already in London. The significance of social class is crucial here, determining the necessity and inclination of a migrant to rely on and to (re)establish his/her transnational network. Relationships among Russian speakers are considered different from those with other Londoners, and the supremacy of the former is explicitly or inexplicitly underlined in migrants' narratives. Friendship represents a specific domain of relationships, occurring on the basis of informal communication and implying emotional attachment and trust. The core element of 'real friendship' is a non-pragmatic, affective reciprocal attitude based upon unconditional trust (Kharkhordin 1999, 2009; Shlapentokh 1989). Friends are a source of moral support, while acquaintances mostly provide 'tangible' help.

Migrants from the FSU, after coming to London, keep their existing understandings of friendship and acquaintance. Interviews show that friendship is commonly related to the issue of Russian-ness. I argue that friendship is significant for the majority of migrants, being a kind of sociality that contributes to the formation of a community not based on kinship or neighbourhood; often, features of transnationalism can be tracked in these relationships. There can be a connection between a small-scale personal relationship like friendship and the formation of a bigger loosely knit migrant community based on the national and cultural origin of its members.

My own Russian-ness informed my relationships with my informants, and blurred the difference between the social roles of interviewer and respondent. After the process of self-representation has been fulfilled, another issue arose – how to build relations with the informants. The problem is one of defining the level of closeness to the research participants, 'an intricate process of identifying spaces and times when it was desirable for me to be an "insider", and situations when it was more desirable to be an "outsider" to the social group under inquiry' (Mullings 1999: 343). Certain strategies described below were employed by my respondents, promoting my inclusion to their groups and making me closer to the insider position.

1. *Categorization of Russian speakers.* Russian-speaking respondents projected their existing taxonomies of possible friends/acquaintances on the researcher. Their dispositions were partly discovered through the way they

treated me – as a representative of a certain category of Russian speakers in London. Marina, for example, tries to keep distance from other Russian women, as encounters with them did not work out well for her:

> The women I've met here…they always need something from you, they have to prepare some documents, have problems… […] Those Russian women you meet here are prostitutes, you know, they just fuck around. Some of them, at least. I try not to socialize with them, they are dishonouring the nation. […] I have my family, I cannot live like they do. Maybe I just do not see nice girls? Well, you generally see less good things in life. Those who stick to you are desperate for something. One of them was asking me to find a man for her, so that she could get married. She was offering me £2,000…

She was less suspicious of me for a number of reasons. I did not resemble the Russian-speaking women who tried to 'befriend' her during the 12 years of her life in the UK:

> I cannot say that all Russian girls are whores. You don't look like one. You are cool. My cousin is a great girl, too. And Sasha is nice and kind. So, I cannot say bad things about everyone…'.
>
> (Marina, 32 years, July 2009)

2. *Mutual activities as a way of distinguishing from non-Russian speakers.* My fieldwork in many cases led to the expansion of the researcher–researched relationship to some non-research-related interactions. It posed ethical quandaries: if what I was doing and talking about with migrants could be counted as research, and whether the information obtained in this way could be counted as ethically pursued and appropriate for academic publications. Oakley (1981: 55) underlines that these problems, generic for qualitative research, are greatest when there is least social distance between the interviewer and the respondent – especially when they belong to the same minority group.

One of the things not approved by the UCL Research Ethics Committee guidelines but frequently turning up in the course of fieldwork was alcohol. While interviewing at respondents' homes, at some point I would notice the respondent getting out a bottle of wine and pouring two glasses. Refusing did not work, sounded impolite, and could make the balance between us more unequal – first, because we would end up being in different states of mind if one of us would be drunk and the other not; secondly, because perceptions of communication while drinking and sober differ in degrees of informality. Engaging in the same practice *with* the informant, however, served as an ice-breaker. Respondents became more relaxed, and the conversation lost some of its formality. Professional roles in this interaction started to fade away, and the respondent perceived the researcher much more as just another Russian than an academic. It gave me a chance to explore the development of a relationship from the inside.

> It's after 1.30 – last call has already passed. There are a few customers in the bar, finishing their drinks. An English guy is sitting at the bar, trying to involve me into a conversation. He wants another drink, but Sergey apologizes and tells him they do not serve any more. I've finished my beer a couple of minutes ago. Sergey asks me:
>
> – One more?
>
> I nod. He gives me another pint. The English guy says in a surprised and a slightly offended tone:
>
> – Hey, but she can have a beer!...
>
> – She already paid for it before the last call. – Sergey replies calmly.
>
> I have not. Moreover, I know he will not ask me to pay.
>
> Sometimes it is alcohol which shows the distribution of power and draws a border between Russian-speaking regulars and others: drinks can be served for free for members of the group, they can be served after the last call and even after the closure. Alcohol is one of the indicators of informal relationships between members of this group – while making a drink for a friend, the Russian-speaking bartender goes beyond his professional role and becomes emotionally involved into this interaction.
>
> (bar ethnography, July 2009)

By getting involved in the studied group's social practices, the researcher becomes initiated into the group member's lifestyle. The ease of such involvement and the disposition of the migrants to socialize with the ethnographer, however, are not universal. Feeling accepted at the bar (where most of the migrants were young and did not occupy high socio-economic positions) was more achievable than making a similar relationship with people 10 years older than me, with higher incomes, married and with families.

3. *Initiative of inquiry.* Another way of making the interaction more informal is for the respondent to take the initiative. Sometimes, informants tried to take control of the research process, expressing their views on the sampling, interview procedure, and analysing the results. Of course, they had their own ways of doing that, presumably more effective than mine.

> R: Yes, considering your research, I can find you a lot of people to speak to.
>
> I: It would be appreciated, actually…
>
> R: How many do you need?
>
> I: Around five people would be nice.
>
> R: Hehe, do you need five *nice* people?
>
> I: No, I am saying it *would be nice* to interview five more. They could be bad, I don't mind.
>
> R: There's no point in interviewing bad people. Some Georgian – what would you need him for? Ha!
>
> (Viktor, 21 years, August 2009)

As well as 'contributing' to the research process like this, respondents commonly take on the role of interviewer, starting to ask questions that interest them. It interrupts my own series of questions, and is at least confusing. It can be regarded as biasing the interviewing practice, but frequently emerges in the course of study and is problematic to avoid. However, the feminist research approach considers it normal to answer the respondents' questions, as it promotes a less exploitative attitude, and builds rapport (Oakley 1981: 47).

4. *Background knowledge.* Asking questions is one of the most common practices of qualitative research. It is at this moment when the researcher's presumably 'insider' positioning strikes back. As a member of the group, you are not supposed to ask questions about most basic, or very specific, rules of behaviour. Being of the same origin, you are supposed to know that. Questions about such 'common knowledge' issues are received with embarrassment. At best, the replies are confined to general descriptions: 'How do you recognize Russian speakers? By appearance, by face...well, *you know all that*'. At worst, people start quoting your work: they have googled before meeting you, giving a condensed account of your own thoughts.

> I: So, who lives in your area?
>
> R: I think you wrote somewhere that in this country Russian speakers do not try to settle together. You know all that. When I came here, I've been asking people where Russians lived in this city. There is no area like Brighton Beach....
>
> (Vladimir, 40 years, October 2009)

Even if the researcher already knows how the things he wants described usually happen – from her own experience or from others' stories – the task of representation implies the necessity of getting the information from the respondents' narratives. The researcher has the responsibility of representing the lives of the studied group to an academic audience, which largely does not share the origin and cultural background of the researcher and migrants. The person doing ethnography finds herself in an ambivalent position, with one foot in academia and another in the sample. Keith (1992: 551) writes about a crisis of representation 'in terms of the relation between subject matter and narrative to the cost of consideration of the relation between representation and audience'. Considering the audience as part of the research process, and the imposition of the authorial interpretation of the 'Other' is another issue of power relations.

I have described how certain features of Russian-speaking migrants' informal socializing interact and counteract with the research practice. The informants' initiatives to erase borders between themselves and the researcher sometimes become manipulative. The inclusion of the researcher by informants included comparing her with the respondent's images of compatriots, engaging in the same activity, practically or discursively distinguishing her from non-Russian speakers, role switching by means of asking her questions, and assuming that she shares background knowledge about the life of migrants in London. All these are also reflections of strategies of inclusion common for socialization of migrants in the Russian-speaking community. I argue that the bigger involvement of the researcher in the studied community, although it can help get access to more sides of migrants' everyday lives, is sometimes problematic for pursuing the aims of the research. The relationship is taken for granted as a friendly one by respondents and sometimes by the researcher. This taken-for-granted-ness may bias the academic representation of the research.

The end

Having briefly described friendships among Russian-speaking migrants, I do not aim to present an idyllic picture of a collective bounded together with affectionate ties. Being a migrant myself did not remove the problems. Differences – socio-economic, educational, generational, gender, legal, country of origin, marital status – between the researcher and the researched will always find their way out and influence their relationships. I've been accused of various faults, from having nothing to talk about because of being 'all about your science and books', to not calling and not coming to a party and therefore 'acting like you are not human'. Russian-ness helped at the beginning, but as the relationships developed, dissimilarities emerged: 'Differences of religion or age or class or occupation work to divide what similarities of race or sex or nationality work to unite' (Merton 1972: 24).

All this made me think about my position among the people I was studying. Abu-Lughod (1991) argues that feminists and 'halfies' – people with mixed ethnic or cultural identity – are the most critical groups for cultural anthropology, able to reconsider the value of the concept of culture by unsettling the boundaries between self and other. At the same time, 'for halfies, the Other is in certain ways the self, there is said to be the danger shared with indigenous anthropologists of identification and the easy slide into subjectivity' (Abu-Lughod 1991: 141). The dilemma of researchers studying their migrant compatriots means positioning themselves in both communities and representing subculture as a way of representing themselves.

Detached involvement (Nash 1963, cited in Agar 1980: 50) is a way of positioning oneself in the course of fieldwork:

> One is, at the same time, part of and distant from the community. One struggles to understand with involvement in the society; at the same time, one stands back critically to examine what one has learned. However, this detached involvement – this stepping into and out of society – is a strain in its own right. There are two obvious ways to lessen the strain. Either keep your distance or 'go native'. You keep your distance at the risk of failing to understand the complexities of a human situation different from your own. You go native, but then stop functioning as a social scientist. Actually, real ethnography represents some of both these strategies as the ethnographer moves around the goal of detached involvement.

The problem of detached involvement is entirely relevant to studying one's own community, especially if it is a minority group. As Agar (1980: 52) puts it, 'while working in your own society, you still have the stress of detached involvement, compounded by the substitution of frequent repeated mini-doses of culture shock in place of the one huge jolt that you usually get in more traditional forms of fieldwork'. Culture shock 'comes from the sudden immersion in the lifeways of a group different from yourself' (Agar 1980: 50) and makes the researcher unable to interpret the things happening around him using his existing knowledge. The confusion can be conditioned by different amounts and combinations of social, economic and cultural capital of the ethnographer and the informant. Some practices are considered normal in one social group but unacceptable in another. Culture shock was hearing from Nastya the story of how she came to London by getting to know a Russian guy from a social website, moving in with him and making him her boyfriend, presented as a suitable way of getting out of Bournemouth. Culture shock was listening to Masha's account of cheating on her husband while loving him and their little son. Culture shock was also desperately trying to find an answer to her question about whether I thought such things should be told to husbands. The inner diversity of the Russian-speaking community showed up again – we behave differently, and nationality did not have much to do with this. Culture shock in this sense was conditioned by differences in our cultural identities that include not only similarities but also 'critical points of deep and significant *difference* which constitute "what we really are"; or rather [...] "what we have become"' (Hall 1990: 225).

Being exposed to a diversity of relationships with different people, and subjugating yourself to their understandings and practices of friendship, puts at risk the researcher's personality. It was even more complicated as I had to fit into different communities. Socio-economic, cultural, generational, gender and other inequalities became more striking, proving that 'the notion of non-exploitative research relations is a utopian ideal that is receding from our grasp' (McDowell 1992: 408). However, whether I succeeded in taking the role of an insider or not, being closer to this migrant group was certainly useful for the research outcomes. I could see from inside the disjunctures that broke this community into stratified groups that this community is highly differentiated, and nationality, ethnic origin or language on their own are not sufficient to bring its members close enough together (Byford 2009).

I recall being particularly happy when I had just finished my fieldwork. It was done! From now on, I thought, I will spend my time writing up, in front of my laptop in my department, surrounded by English books and socializing with my English-speaking colleagues (among whom, I gladly admitted, there was not a single native Russian speaker). I wanted to draw the line between the society I've been studying and the society I live in. It was time to detach myself from drug-taking youth and clean kids from good families, mumsies and women of relaxed morals, girls in their mid-twenties who asked me what fascism was and why the Second World War started and Russian-Jewish rappers that turned out to be fascist, migrants who were reluctant to talk to me, and those who were looking for my company and were less than appealing to me. Friendship seemed a nice thing to write about; however, experiencing many different friendships personally was overwhelming.

In the final part of this chapter, I pointed out how the differences between me and my respondents disrupted or put an end to our relationships, and demonstrated that ethnic/national origin does not provide enough ground for a continuous relationship. Most of the relationships started because of the research; most of them ended not only because the research came to an end but also because of personal issues. This research showed that studying migrants by a migrant researcher poses questions similar to those which arise in feminist geography. Regarding feminine psychology as that of a subordinate group is paralleled with the perception of ethnic minorities (Oakley 1981). How can the mutual positioning of the ethnographer and the respondents be conditioned by their similar origin and status of a national/ethnic minority encountering stereotyped and oppressive attitudes? Even if we consider it as an advantage to (arguably) belong to the same community, it is necessary to look for appropriate ways of representing the life of this community to the academic audience.

My central argument is that belonging to the same community can be helpful in conducting research; however, the diversity of this community should be an object of constant reflexivity, and the seemingly easy access to a group should never prove that you are an insider. I support the feminist argument of making the pitfalls and gaps of a research visible, and claim that practical downsides of a study can turn into or confirm its actual outcomes. By trying to befriend some of my respondents, I was breaking the unwritten rule typical for some migrants, one of the outcomes of my study of informal relationships – the reluctance to make new migrant friends in a stratified community. By trying to detach myself from others, I was breaking another unwritten rule – that of easily established informal relationships and taken-for-granted Russian friendship. But by learning, succeeding and failing to approach different people in the context of personal relationships, I also learned a great deal about how these migrants considered the values of friendship, its social differences, and distinctions between the degrees of closeness.

Acknowledgements

I would like to thank Dr Alan Latham, Dr James Kneale and Dr Russell Hitchings for their guidance and useful comments, Dr Anton Shekhovtsov for moral support, and Dr Ulrike Ziemer for giving me the opportunity of this publication. I am also grateful to the Berendel Foundation for the financial support of this research.

Notes

1 The research was conducted in London, 2009–2011. Ethical approval was granted by UCL Research Ethics Committee and all ethical guidelines were followed in the study. All respondents' names mentioned are changed.

2 In the sample of 35 Russian-speaking migrants, we have 22 women and 13 men, aged 20–45, with the biggest proportion of those in their late 20s. All of them are first-generation migrants, having lived here from two to 12 years; 24 are Russian (ethnically, three of those are Jewish, one is half Tatar, one is half Jordanian); five are from Ukraine; three from Belarus; two from Latvia; one from Lithuania; and one from Tajikistan. Some of the informants have previously lived in other countries such as Israel, South Africa, Spain, Scotland, USA, Canada, Germany, Italy, Hungary, Switzerland, Saudi Arabia, Greece and Jordan. Three of the informants are illegal migrants, living on fake passports: one is a bartender, another does occasional temporary jobs (repair, transportation, security) and the third is a criminal. The other participants have occupations that include a musician, two bartenders, an interior designer, four PhD students, an elite property consultant, a nanny, three housewives, a City analyst, a rapper, a scholarships administrator in an international charity organization, six Bachelor students, a shop assistant, an executive assistant at one of London schools, a driver from a car hire service, a phlebotomist, a risks analyst, an interior designer, a person working for human rights organization, an IT specialist, and a couple working in consulting and marketing research and tourism-related property consultancy. I deem as a valuable feature of this research that it is not limited to any relatively homogeneous social group.

8 Facets of migrant identity

Ethical dilemmas in research among Romanian migrants in the UK

Oana Romocea

The present chapter focuses on the ethical issues arising from conducting qualitative research among Romanian migrants settled permanently or temporarily in the United Kingdom (UK). It has a two-fold aim. First, it aims to be analytical by investigating the dilemmas faced by researchers in their empirical fieldwork. Secondly, it also has a self-reflective dimension that is prompted by the fact that I, as the researcher who is also a member of the Romanian community in the UK, identified myself in one capacity or another with the interview participants. By considering and acknowledging these two aspects before starting the research process, I attempt to avoid being either ethically neutral or morally prejudiced. The ethical neutrality is a stance which detaches the researcher from the participants' experiences (Iosifides 2011: 216). As a consequence, this might limit the researcher's understanding of the participants' daily routines and practices in a migration context and prevent the researcher from showing 'genuine concern about participants' social situations, perspectives and actions' (Iosifides 2011: 215). In contrast, by adopting a position of moral relativism, the researcher's main aim is to promote a dominant discourse and a predefined world-view rather than approach the research project with an open mind and allow room to be surprised by the research findings. This stance compromises the research by not allowing room for self-reflexivity and progressiveness on the researcher's part.

The chapter addresses the ethical stances which occur during the different stages of the research process: from the early stages of formulating the research questions, to identifying the key elements of the research design, to the data-collecting phase and ending with the delivery of information. It is essential that one of the best practices employed by researchers today is to reveal and explain the research methods and any ethical considerations related to the research project (Jacobsen and Landau 2003: 2). By making known these details, their research meets the dual imperative of being both academically rigorous and policy relevant (ibid.: 1).

Migration-related research entails a growing set of ethical implications to be considered by the researcher at each study stage due to the nature of the subject, which 'is not morally neutral' (Birman 2006: 164) but rather encumbered by cultural sensitivities, different world-views, distinct social and moral values and intense public discourses. Overall, this chapter aims to contribute to the literature on the ethical considerations of conducting research within migrant communities as well as offering insight into ethical best practices. Throughout this chapter, the ethical claims are given a meaningful interpretation by connecting them with concrete examples from my own research conducted with Romanians settled permanently or temporarily in the UK.

The chapter is organized in three main sections. First, I offer a description of the Romanian migrant community settled in the UK by looking at its origins and producing a typology of Romanian migrants in the UK. Secondly, this chapter highlights both the advantages and the challenges of doing qualitative research among migrants. The third section addresses a selection of ethical considerations which arise from doing research within migrant communities. This chapter concludes that the researcher needs to develop a common-sense approach based on balancing the potential of the research to bring positive and negative benefits and on a negotiation process between detached objectivity and compromising subjectivity.

Setting the context: Romanians in the UK

After the fall of communism in 1989, Romania joined with the other countries of Eastern Europe in their pursuit of social liberty, political freedom and economic recovery after 45 years of an oppressive political regime and a state-governed economy. A long-nurtured aspiration to become more like Western Europe suddenly appeared achievable. Their ultimate identification with Western Europe was the membership of the two international organizations: the North Atlantic Treaty Organization (NATO) and the European Union (EU).

NATO welcomed Romania as a country member in 2004. However, its accession to the EU suffered a drawback in 2004 after its failed attempt to join the EU in the same enlargement wave with the other 10 Eastern European countries. Eventually, Romania officially became a full member three years later, on 1 January 2007.

Romania's accession to the European Union opened the door for once-in-a-generation opportunities for Romanian citizens, such as visa-free travel and free movement across Europe, access to quality higher education at reasonable tuition fees and access to the Western labour market. Since 2007, there has been a steady exodus of Romanians from all walks of life towards Western Europe. Romanian students are present in most major university centres. As for the labour force, although there are no official data available, it is estimated that over two million Romanians, representing 10 per cent of the country's population, are currently working outside the Romanian borders, the vast majority in Western Europe (Erdei 2008). This migration has triggered the emergence of numerous Romanian migrant communities in various EU countries.

In a poll conducted at the beginning of 2007, the UK was ranked fourth in Romanians' preferred destination countries for migration, after Italy, Spain and Germany. Before Romania's accession to the EU, the number of Romanians living in the UK was relatively small, reportedly around 7,500 according to the 1991 UK Census (BBC 2001). During the 1990s, there was a low but steady flow of Romanians relocating to the UK; from 2007 onwards, a larger migration wave has been taking place. A report in the *The Observer* mentioned that around 50,000 Romanian workers had relocated to the UK in the first months after Romania's accession to the EU in 2007 (Helm 2008). Recent estimates revise this figure to around 83,000 Romanian-born citizens currently living in the UK (Office for National Statistics, March 2011), while the Romanian ambassador in the UK has recently stated in a speech given in July 2012 that the Romanian community numbers approximately 130,000 people (Jinga 2012).

Current migration scholarship has called for a methodological shift to study such sizeable transborder flows. Until recently, the nation-state provided the predominant analytical framework for understanding contemporary migration movements across political borders (Georgiou 2007; Glick Schiller 2008; Glick Schiller and Levitt 2004). This macro-level approach facilitates research from the perspective of policy makers (Bilger and van Liempt 2009: 2) and reveals migration patterns through figures and statistics in which 'flows replace individuals and the motive of migration in a simplistic and generalized way to a point where they have little significance' (Apap 2000: 196). Nevertheless, it proves to have its shortfalls, especially in terms of engaging with the micro-level, everyday life of the migrant and in understanding how daily experiences affect migrant identity and outlook in the diaspora. To remedy the problem of overlooking experience, I embrace ethnographic methods in my research that have allowed me to collect in-depth information about the process of migrant identity formation. Generally speaking, ethnographic enquiry seeks to reveal the participant's perceptions and understandings in the context of their overall world-view (Crotty 1998: 7). Ethnography also allows researchers to document, study and understand how people interact with both their host society and homeland and how these links affect their daily experiences as migrants (Glick Schiller and Levitt 2004: 1013).

For the current study, I rely on data collected through in-depth interviews that bring depth to our understanding of the migration experience of Romanians settled permanently or temporarily in the UK and the formation of their new identity as migrants. One of the main characteristics of the Romanian migrant community is its diversity generated by variables such as education, age, time of uprooting, the emigration experience, access to existing social networks, access to technology, etc., which can potentially influence their migrant experience. Given the significant number of variables, one may argue that any qualitative small-scale research would not be representative of the whole Romanian migrant community living in the UK (Jacobsen and Landau 2003). Nevertheless, I believe that if the researcher describes in detail the key elements of the research design, any issues regarding the validity and the representativeness of the research can be addressed. This is what I aim to do in this present chapter.

For my current study, I classified the Romanian migrant community into three main groups based both on the time their migration took place and on the reason of their migration: pre-1989 diaspora; post-1989 academics; and post-1989 labour migrants. I tried to randomly select participants from these three groups in order for the sample group to be as representative of the migration experience as possible. This is a key ethical issue, as Birman points out:

> An ethical approach to research is to insist on including the diversity of immigrant and refugee populations in research and to include questions about the variety within-group variations in the research protocol.
>
> (Birman 2006: 164)

Although there are differences between the three main migrant groups considered for this study, they also share various characteristics. On the one hand, all the interviewees are first-generation migrants. This aspect is crucial, as all subjects lived the same transitional experience of uprooting from their home country and of adjusting to the host culture. A comparative study of the three main groups forming the majority of the Romanian migrants settled in the UK will enable us to understand the complexity of the ethical issues in research with migrant communities. This approach reveals a two-dimensional analysis of the Romanian migrant community: a longitudinal cross-section which studies how the migration experience is dynamic and evolving in time and a latitudinal or spatial cross-section which looks at the migrant community in the current economic, political and social climate. Very few studies seem to have made a genuine attempt to compare different places and time periods (Apap 2000: 196).

The pre-1989 category consists of political refugees and can be described in Cohen's terms as 'victim diaspora' (Cohen 2008: 2). They left Romania during the communist regime and came to the UK to ask for political asylum. Most of them were educated people whose main reason for emigration was ideological in nature. Upon arrival to the UK, they did have access to an existing, although small, diasporic social network but chose to be or not to be part of it. There are various reasons for each choice. The individuals who avoided any contact with other Romanian co-nationals did so because of fear that they might be reported to authorities and then deported back to their own country where they would suffer cruel consequences and persecution. The other reason for their lack of involvement with the then-existent Romanian community is the lack of trust towards fellow Romanian co-nationals. The migrants who chose to get in contact with other Romanians already living in the UK did it for different reasons, typically to seek help, socialize, organize political opposition against the communist regime in Romania or to get news from home, given the difficulty of accessing reliable information.

Their status changed over the years. If they entered the UK as political asylum seekers, after 1989 they were identified as migrants. In other words, before 1989, they belonged to the Romanian diaspora in the UK, but after 1989 they were identified as migrants. Their status change is also reflected by the way their transnational connections developed, adapted and transformed over the years. At the time of their arrival and immediately afterwards, they intentionally avoided establishing any links with family members who they had left behind, as they did not want to jeopardize their safety, given the political and social oppression that existed in Romania before 1989. Aurica, a lady who left Romania in the 1970s when she was only 30 years old, describes the situation in metaphoric terms:

> When one leaves the country legally it is like one is crossing a bridge. Once you are on the other side, the bridge is still there. If one left the country illegally, as both my sister and I did, after you crossed the bridge, you have to destroy the bridge. We knew that behind us the door must be closed and locked. We could not get back and we had to accept it in our hearts and our minds that we had to build a new life here.

If immediately after their arrival in the UK, the pre-1989 migrants had no connections with the homeland, their transnational connections changed once the political and social setting in Romania changed. The same lady describes these circumstances:

> After 89, we immediately began rebuilding the bridges with family and friends back home. I invited all my friends to come over. They stayed for a month, I took them everywhere.

The two post-1989 categories moved to the UK with different ambitions. The main aim of the knowledge diaspora, i.e. academic-related migrant groups and highly skilled migrants, is personal development by gaining qualifications and opening up career perspectives. When they arrive in the UK, they already have a good knowledge of the host society provided either via the academic institution attended by the academic-related migrants or by the recruitment agencies dealing with highly skilled migrants. They also have a good knowledge of spoken and written English which facilitates their direct and immediate contact with the host culture. Their status provides a degree of security and predictability: financial, accommodation, introduction to the host culture. Most of them are in their 20s or early 30s, without dependents. If they have a family, it is likely that the family accompanies them. For this group of migrants, the existence of a transnational social formation is not essential. Augustin, a student who came to study in Bristol in 2007, was impressed how smoothly his cultural transition took place:

> When I arrived in the UK, everything was fine. I arrived in Bristol in the morning and by noon I was already given the key to my dorm room. I was pleasantly surprised at how fast everything happened. No problems whatsoever.

The economic migrants' goal is financial gain. They fall under Cohen's category of labour diaspora (Cohen: 2008: 61). Their arrival in Britain is shadowed by the uncertainty of everyday life: i.e. finding a place to work and live. Their lives are characterized by unpredictability. Although in many cases, their relocation is often facilitated by kinship or friendship networks, the work migrants come to the UK uninformed about the host culture. The information they receive from friends and relatives is fragmented and often does not reflect the whole reality. The aim of their relocation is the promise of a better life, good earnings and rapid betterment of their economic status. However, upon their arrival in the UK, most of the labour migrants soon face disappointment as the dream and expectations they had prior to the relocation do not match the reality. Establishing contact with an existent transnational social formation is crucial, as it helps them fulfil the aim of their relocation: i.e. finding a job and a place to live. Hence, they have strong links with the migrant community and very limited interaction with the host society and culture. Often the language barrier is also a hindrance, as most of the labour migrants do not possess knowledge of conversational English at the time of arrival in the UK. They are more likely to be alone, leaving their families behind in Romania. These circumstances are reflected by Paul, a young builder who arrived in the UK to work for a building company owned by a Romanian co-national:

> I arrived in the UK with the help of an acquaintance. I came to work in the building trade in a Romanian company. I didn't have too many connections with British citizens in the beginning. I was living in a house with another 5–6 Romanians. The company offices and the building site were all in the same part of London.

In terms of geographic distribution, the Romanian migrants are spread across the whole of the UK, forming clusters around big cities such as London, Bristol, Birmingham, Leeds, Glasgow, Cardiff and Belfast. This is reflected in the number of Romanian churches in the UK established after 1989 (12 in London, one in Birmingham, one in Leeds, two in Bristol, one in Poole, one in Liverpool, one in Glasgow and two in Northern Ireland). These have provided a gravitating centre especially for the Romanian economic migrants.

The differences between the three groups raise important ethical questions and implications, some of which I analyse here. These include access to the community, the relationship between the researcher and the participant, the concept of trust, the legal component of the fieldwork, emotional involvement, the sensitivity of the questions asked, cultural sensitivity and the impact of the research on the participant. This list is not exhaustive. Although all these ethical aspects were present in my dealings with the three diasporic and migrant groups, I discovered over the course of my fieldwork that different ethical issues were more specific and more evident in one particular migrant group of the three examined. For example, the legal component of the fieldwork (Düvell *et al.* 2008: 5) featured more prominently when dealing with the labour migrants, whereas the emotional dimension (ibid.: 6) had greater relevance when speaking with Romanians who came as political refugees before 1989.

It is worth mentioning that there are at least two other identifiable distinct categories of Romanian migrants: the Romanians of Roma ethnicity and the Romanians who arrived in the UK via illicit means such as trafficking and smuggling. These two fringe migrant groups deserve separate studies, but they did not constitute the focus of this research for other reasons. Doing research among these migrant groups raises another set of ethical dilemmas, including safety issues on the part of the researcher (Düvell *et al.* 2008: 13). These ethical concerns are more complex in nature and require separate attention, but as these groups were not included in the research project in question they are not discussed in this chapter.

Research methodology and sources

I consider ethnography an appropriate and instrumental methodology in collecting in-depth information about the process of identity formation among migrant groups. Ethnographic enquiry seeks 'to uncover meanings and perceptions on the part of the people participating in research, viewing these understandings against the backdrop of people's overall worldview or culture' (Crotty 1998: 7). Rodgers uses an informal term, i.e. 'hanging-out', to define this methodology. Despite using non-academically standard terminology, he insists that this kind of qualitative research is 'an indispensable research tool that is essential for the formulation of informed, creative and self-critical responses' (Rodgers 2004: 49). It allows the researcher to discover the 'meaning making processes and interpretative understanding of the social world' (Iosifides 2011: 1).

Ethnography gives the researcher the opportunity to reveal the subtleties of the migration process, such as causes and effects of migration, migration decision-making processes, identity formation, the role of social capital and social networks and the dynamics of transnational social spaces. Conducting qualitative research at the micro level is essential in understanding how 'political struggles of everyday life are linked to relationships and processes of global significance' (Rodgers 2004: 49). Analysis at the micro level imbues the research with a humanized and personal dimension.

The call for a methodological shift to study transborder flows come from a number of scholars. The social scientists Glick Schiller and Levitt (2004: 1012) argue that 'we need a methodology that allows us to move beyond the binaries, such as homeland/new land, citizen/non-citizen, migrant/non-migrant, and acculturation/cultural persistence that have typified migration research in the past.' They believe that ethnography allows researchers 'to document how persons simultaneously maintain and shed cultural repertoires and identities, interact within a location and across its boundaries, and act in ways that are in concert with or contradict their values over time' (ibid.: 1013). Bird also acknowledges the unique advantages of ethnography, believing that 'the on-the-ground perspective of the ethnographer is still crucial and offers a dimension that no other approach can duplicate' (Bird 2003: 190).

Reflecting on the new methodological approach, this current research aims to disclose the daily routine of practices through research at the grass-roots level. In other words, it tries to understand its research subjects in the context of their real lives and experiences. By employing ethnographic methodology, this research also aims to distance itself from the predominant approach in migration studies which uses the nation-state as an analytical perspective (Georgiou 2007: 26) whose main shortfall is its inability to engage with everyday life and to reveal the implications of how the migrant status is framing the daily experiences. On the contrary, doing fieldwork is the most sensitive aspect, as it brings the researcher in direct contact with migrants and their environment.

For the purpose of this study, I conducted 33 interviews with Romanians who are currently settled in the UK. They are representatives of the three migrant groups described above. The age of the interviewees ranges from early 20s to early 80s and they are mostly settled in England, the majority living in London and its surrounding areas. I aimed to include as diverse a body of participants as possible. The way I identified them was by the recommendation technique, i.e. background links established during respondent discussions. I identified a number of migrants in each group and approached them for interviews. After I conducted the interviews, I asked the participants to enquire among their Romanian friends if there were any willing to participate in the research. Once I received a positive reply, I approached the new participants either by phone or email. This technique helped me in the selection process, as I was able to establish a mental profile of prospective participants before I approached them. Hence, I was able to assess their suitability for my project. Following this technique, I felt I was in control and managed the sample of migrants I interviewed.

The interviews were conducted mainly in Romanian with two exceptions – two young professionals, who claimed that English was their first language, owing to the fact that they needed to use English on a daily basis, one being married to a European national and one to a British citizen. However, the interviews ended up being bilingual, depending on the experiences the two interviewees were recounting. Knowing both languages, i.e. the language of the migrant community and the language of the host country, was a key advantage while conducting this research since it gave the interviewees the opportunity to express themselves in whichever language they found more comfortable and I, as the researcher, could interact with them.

The interviews took place in various places, including public spaces such as cafes, churches, offices and private homes. The pre-1989 migrants mostly preferred to give the interviews in their homes, often in a social context, such as a meal or over non-research-related discussions. Thus, the whole interviewing experience was transformed into a social event. They spoke at length about their pre- and post-migration experiences, often revealing facts before or after the recording session. I recorded this information separately in a notebook.

Given their diverse backgrounds, experiences and ages, the interviews required a great deal of adaptability and flexibility on the part of the researcher, although the qualitative method of in-depth interviewing did provide flexibility. This attribute of the qualitative interview, which allows the researchers to adapt it to the various circumstances of the research setting, leaves room for choice and it is within this context of choice that ethical issues come into play.

The relationship between researcher and interviewees

Within the flexible framework of qualitative research, a complex relationship between the researcher and the research subjects is established. The very nature of this relationship raises important ethical considerations that I will debate below. Although all the ethical aspects which I discuss here were present in my dealings with the three migrant groups, I discovered that different ethical issues were more specific or more evident in certain migrant groups compared to others.

First, I consider the issue of trust. In order for the interview to reveal information relevant to the research, the relationship between the researcher and the interview participant needs to be based on trust. A key element of this trust-building relationship is effective communication before, during and after the interview.

As a researcher, I made an effort to gain the prospective participant's trust from the moment we had established contact. I was very open and transparent about the intentions and purposes of my research. The first step was to present them with a short abstract of my project. The aim of the abstract was three-fold: first, it informed them of the framework of the research and, consequently, the direction of the interview and the kind of information I was looking for; its second aim was to 'sell' the research project to the participant by raising their interest in the subject; and thirdly, it created a platform for two-way communication between the researcher and the participants where any concerns could be raised in the initial stages of the research. I also provided them with a brief biography so they would know who would be interviewing them.

The information I provided in the first instance proved to be a great platform for when we eventually met for the interview. The terms of the interview were established in advance and I explained to them that their personal details would not be revealed. The guarantee of anonymity seemed to be essential for both the political diaspora and the post-1989 labour migrants, but for different reasons. It did not seem to raise any worries with the younger generation of knowledge diaspora. This entire process is known in the field of research ethics as informed consent. The current standard research protocols established in Western institutions require the participants to sign an informed consent form. Although this procedure has been established with good intentions, in some cases this can instil fear and concern in the participants. In some cultures, such as the Romanian one, this practice is often viewed with suspicion. In my case, most of the interviewees were open and willing to give the interview as they were aware of a 'certain British manner' in dealing with these issues. However, at the end, when I asked them to sign the form, I often noticed a surprised reaction.

This was mostly due to the fact that Romanian culture is not a culture where one's rights are protected by legally binding forms. It is a culture based on willingness, openness and common sense rather than suspicion and negative anticipation. For most of the participants, the fact that they agreed to give the interview should, in their opinion, suffice as an official form of consent.

The issue of trust is also inherently linked to the issue of confidentiality. Although in the context of the in-depth interview, personal or sensitive data is often disclosed, it is my ethical responsibility as a researcher not to include this information in any output, not least because this was part of the terms and conditions of the interview in the first place. Gaining trust is indispensable for doing research. The trust I built with individual interviewees advanced my research, as I found most of my participants via word of mouth, based on the recommendation received from previous interviewees.

Why is the issue of trust so important? I discovered that it was imperative to gain the interviewees' trust in the first place in order for them to open up and give authentic answers. There is a danger that participants give answers to questions in line with an anticipation of what the researcher expects. Another instance is the 'collective memory' (Appadurai and Breckenridge 1989: i) of the migrants, which means there are certain narratives which are embedded in the migrant consciousness but which do not necessarily reflect their own experiences. One such example was the discrepancy between what the migrants thought about British perceptions of Romanian migrants and what they had experienced in terms of their interaction with British people. The interviewees' opinions were overwhelmingly negative on the subject of British perceptions of Romanians, but this was because their answers were based on media discourse. However, when asked about their own experiences, they admitted that they had not experienced ethnic discrimination to any significant degree.

Another ethical aspect crucial in building trust was the ability to overcome cultural sensitivities owing to the researcher's status as a deterritorialized Romanian national. I am a 'migrant European researcher', a term coined by the network of Young European Social Researchers. They identify academic migrants as individuals who move in another country where they need to use a foreign language for both conducting research and for daily social encounters. They actually live the same uprooting and re-grounding experience as the research participants (Borkert and De Tona 2006). Being a migrant European researcher doing research within my own migrant community put me in the position where I had to draw on both the emic and etic perspectives, permanently negotiating between the participant's view and the observer's view (Cohen 2008: 5).

Familiarity with the cultural context of the participants is a key aspect to be considered when doing research among migrants, as it gives the researchers an emic perspective. Birman suggests that the cultural differences between the researcher and his interviewees and the researcher's lack of familiarity with the participants' culture can pose a number of challenges throughout the research process (Birman 2006: 164). He even suggests that if this lack of familiarity exists, a third party which mediates the contact between the researcher and the participants needs to be considered. However, the use of a mediating research assistant from the same culture as the respondents can without a doubt influence the quality of data collected as it 'risks transgressing political, social, or economic fault-lines of which the researcher may not be aware' (Jacobsen and Landau: 2003: 10).

Being a member of the Romanian migrant community helped me overcome this challenge and I benefited from an insider's perspective. For my interviewees, I was perceived as one of them. As Borkert and De Tona (2006: 6) astutely observe:

> When we go to the field as transnational researcher and we research other migrants, our reciprocal foreignness and otherness can function as a meaningful site of encounter and a way to negotiate closeness.

I could understand their migrant experience, having experienced re-rooting from my own culture and having to adjust to a different host society. Hence my interview questions were specific and not general. Given my inside knowledge, the participants did not hesitate to share the negative incidents or experiences they had encountered while adjusting and integrating in the host society, thus giving me the opportunity to collect valuable authentic, unfabricated data. This might not be the case if the research had been conducted by a British researcher.

At the opposite end of the spectrum of ethical concerns, my status as a member of the Romanian migrant community had a direct impact and influence on my motives and approach of the project, and this, in turn, may have resulted in morally compromised research, polarized research findings, a lack of objectivity and the promotion of a predefined world-view. As Bolak suggests, there is a risk in both being and not being too familiar with the participants' original context. The danger is that 'indigenous researchers run the risk of being blinded by the familiar', whereas the 'foreign researcher runs the risk of being culturally blind' (Bolak 1997: 97). My status shaped the way data was selected, presented and analysed. My view is that as long as I acknowledge my own position as a researcher and the bias that may characterize my approach throughout the research stages, my research has validity. Nevertheless, the whole research process is a continuing negotiation between ethical neutrality and moral relativism (Iosefides 2011: 216).

The other ethical aspect was the migrant status of some of the participants and their conformity with British law. Given the restrictions to the British work market imposed on Romanians, there were cases when migrants flouted British immigration law. This is especially the case in the period immediately after the migrant's arrival in the UK.

As I have already mentioned above, most of the working migrants face a certain set of uncertainties when they arrive in the UK, notably with their future employment and accommodation. Some of them may not have the right documents and some may find themselves living in poor accommodation. Understandably, their status might have prevented them from taking part in the research. However, I believe that the fact that I was a Romanian national gave them reassurance and confidence. Nevertheless, the interviews exposed them to a certain degree of vulnerability and risk. Opening up to a stranger and relating their story and circumstances to the interviewer might compromise the safety of their status as a consequence of their involvement in research. On the other hand, as a researcher, the main ethical dilemma I faced was how to respond to the complex set of legal circumstances specific to their status.

This legal component of the relationship between the researcher and the subject brings us to the next ethical consideration: emotional involvement. As a researcher, how much should I become emotionally involved in the subject's personal circumstances? This aspect was more evident when I dealt with labour migrants and the pre-1989 group. In the case of the labour migrants, their vulnerability in the UK gave me the potential opportunity to get involved and try to help by giving advice related to their particular circumstances.

For the pre-1989 community, the emotional aspect had a different facet. These people were already well-established UK citizens, but the retelling of their migrant experience, although it happened a long time ago, provoked emotional responses. If in the case of the working migrants the emotional aspect was related to their adjusting to another culture and context, in the case of the pre-1989 group, the emotions were triggered when the respondents recounted their departure, which in most cases was traumatic. When one of the interviewees took the decision to leave Romania in the 1970s, she did not tell her family for fear that they would be persecuted by the Romanian authorities if they found out she had left. Another participant told me how she had to leave her husband and baby behind and they were reunited only a year later.

Their stories were often heartbreaking and, when retold, brought up a lot of unpleasant memories. As a researcher, I was very careful in handling personal data. In one case, I had to interrupt the interview and give the subject the opportunity to recompose herself. Although the stories triggered a lot of emotions for the interviewees, in general they were happy to share their experiences with me as they felt that someone is listening to them. I also have to mention that the interviews with the pre-1989 migrants necessitated the most emotional involvement from my side. In most cases, the interview turned out to be a whole social experience and a day out with the interviewee. Although I initially asked them just for up to an hour of their time, they were more than happy to spend most of the day with me, either inviting me to stay for lunch or taking me sightseeing in their city.

The other ethical aspect of my research is related to the concept of sensitivity. There are two types of sensitivity (Düvell *et al.* 2008: 10). On the one hand, there is the sensitivity of my questions and how intrusive I can be as a researcher in order to collect the necessary data and find the answers I need, which entails a level of vulnerability of the participant. This aspect was made evident during my interviews with irregular labour migrants when critical information about their practices was discussed. The public disclosure of these facts through publication of the research findings can cause harm, damage and emotional stress to participants. The ethical implications consist of a careful examination and of the nature of information which is disclosed, omitting specific details in order to protect the research participants (Iosefides 2011: 217).

On the other hand, there is sensitivity related to public opinion and the cultural and political context of the study. Being a part of the Romanian community in the UK, I could better understand the migrants' context of origin which influenced and shaped their definition of what is ethical. However, it did not necessarily mean I agreed with or adopted their views and perspective. I was just able to grasp the full picture of their background without passing judgement.

Delivery of data

Before concluding this chapter, I would like to briefly note some ethical considerations in respect to the presentation of data. One key aspect in data delivery is the process of selecting or filtering the information. There are two selection stages that pose their own ethical questions. First, the participants convey information which they assume is significant. This information might be important for them for emotional or personal reasons, but they might not necessarily bring any contribution to the research project. Often the in-depth interviews convey more data than the researcher needs. The researcher's responsibility is to try and keep the interview focused by asking questions which bring the discussion back to the subject matter.

The second selection process is the researchers' interpretation of the collected data and the way they present and disseminate it by means of articles, books, conference presentations, posters, etc. At this stage, the ethical consideration involves the appropriate portrayal of the participants in the research project. When it comes to the presentation of data, it is the researcher's task to build each participant's narrative and past experiences relevant to the research project by whittling down the interview to the essential data (Rosenthal 1993: 62).

One of the ethical dilemmas I encountered during my research with the three types of Romanian communities in the UK was the use of appropriate terminology. The word 'migrant' is largely used in the policy documents and media. Because of the way it is used in these fields, it has a negative connotation which is mostly associated with people who relocate to another country in order to take advantage of the opportunities it offers. As Düvell *et al.* rightly highlight,

> technical legal or political expressions or political jargon find their way into academic literature; this practice even implies that terms and concepts are accepted and their meaning taken for granted. Much of these terminologies, however, serve a certain purpose, such as administering migration, facilitating certain policy processes or manipulating public discourses and are accordingly loaded with ideology and politics (Düvell *et al.* 2008: 24).

In legal terms, all the interviewees were migrants, but when I asked them during the interviews if they considered themselves migrants their answers varied. The labour migrants readily assumed this status. However, the participants from the other two groups – the pre-1989 political diaspora and the knowledge diaspora – had difficulty identifying themselves as migrants although they admitted their migrant status eventually. Therefore, in this chapter, I tend to use the word 'migrant' mostly when referring to the former group and 'diaspora' when talking about the latter two groups.

It is always difficult for a researcher to know or foresee how the research content which has been made public via either published work or conference presentations will be used or interpreted by a third party. Therefore, the researcher's responsibility is to ethically judge the content presented (ibid.: 24). While collecting data, the researcher's main responsibility is to maintain ethical standards at the micro level by sourcing and representing someone else's narrative and history in an accurate manner. At the stage of data dissemination, the ethical implications need to be considered at the macro level. Researchers have a responsibility to make sure their research is used to inform correctly the 'policies that have great impact on the lives of many people' (Birman 2006: 155).

Conclusion

I have presented different types of ethical questions and implications involved in migration research which transpire mainly from the relationship established between the researcher and the participants in the research project. These include the concept of trust, the legal component of the fieldwork, the emotional involvement, the sensitivity of the questions and the cultural sensitivity. If ethical issues at the data-collection stage seem evident, then I have shown that ethical aspects may also appear at the later stages of the research project, including the delivery of data. Regardless of the research stage, ethical considerations are complex in nature and researchers need to develop 'a sophisticated understanding of the underlying issues so that they can negotiate creative solutions to resolve them' (Birman 2006: 156).

I have shown that the ethical implications of doing research with migrants are significant on both the micro and the macro level. Therefore, several conclusions can be drawn. First, the researcher should always be mindful of the well-being of the participants, both while collecting data and once the research results are made public. Secondly, researchers are not free of bias. Instead, researchers need to acknowledge this fact and discuss openly their views and perspectives. This will validate their research, which will contribute to the academic debate in the field and allow for 'engagement in genuinely critical scientific inquiry' (Iosifides 2011: 219). Thirdly, the research agenda, aims, findings and conclusions need to be clearly stated so there is no room for speculation when the research findings are used and interpreted by policy makers or third parties. The main conclusion is that doing research among migrants requires a constant balance between the potential to harm and the benefit the research findings will bring to both the migrant community and the host society. Ultimately, this ensures the scientific standard of the research.

Given the complexity and multiplicity of ethical considerations, this chapter did not plan to be exhaustive. It merely exposes some specific challenges which were faced in the given circumstances of my research. Research of different ethnic migrant groups within the British context or within Romanian migrant groups in other countries might expose the researcher to different ethical dilemmas.

References

Abramowicz, B. (1979) *Ze wspomnień rodowitego berlińczyka*. Zielona Góra: Lubuskie Towarzystwo Naukowe.

Abu-Lughod, L. (1991) 'Writing against Culture', in R. Fox (ed.), *Recapturing Anthropology: Working in the Present*. Santa Fe, NM: School of American Research Press, pp. 137–62.

Adamson, F. and M. Demetriou (2007) 'Remapping the Boundaries of "State" and "National Identity": Incorporating Diasporas into IR Theorizing', *European Journal of International Relations*, 13(4): 489–526.

Agar, M. (1980) *The Professional Stranger: An Informal Introduction to Ethnography*. New York and London: Academic Press.

Agnew, J.A. (1994) 'The Territorial Trap: The Geographical Assumptions of International Relations Theory', *Review of International Political Economy*, 1: 53–80.

Agnew, J.A. (2009) *Globalization and Sovereignty*. Lanham, MD: Rowman & Littlefield Publishers.

Albert, M., D. Jacobson and Y. Lapid (2001) *Identities, Borders, Orders: Rethinking International Relations Theory*, Borderlines, Vol. 18. Minneapolis, MN: University of Minnesota Press.

Ali, S. (2003) *Mixed-Race, Post-Race: Gender, New Ethnicities and Cultural Practices*, Oxford: Berg.

Amin, A. (2004) 'Multi-Ethnicity and the Idea of Europe', *Theory, Culture & Society*, 21(2): 1–24.

Anderson, B. (1992) *Long-Distance Nationalism: World Capitalism and the Rise of Identity Politic*. Amsterdam: Center for Asian Studies.

Anthias, F. (1998) 'Evaluating "Diaspora": Beyond Ethnicity?', *Sociology*, 32(3): 557–80.

Apap, J. (2000) 'Investigating Legal Labour Migration from the Maghreb', in B. Agozino (ed.), *Theoretical and Methodological Issues in Migration Research*. Aldershot: Ashgate.

Appadurai, A. (1991) 'Global Ethnoscapes: Notes and Queries for a Transnational Anthropology', in R. Fox (ed.), *Recapturing Anthropology: Working in the Present*. Santa Fe, NM: School of American Research Press, pp. 191–210.

Appadurai, A. (1996) *Modernity at Large: Cultural Dimensions of Globalization*. Minneapolis, MN: University of Minnesota Press.

Appadurai, A. and C. Breckenridge (1989) 'On Moving Targets', *Public Culture*, 2: i–iv.

Appiah, K.A. (1997) 'Cosmopolitan Patriot', *Critical Inquiry*, 23(3): 617–39.

Appiah, K.A. (1998) 'Cosmopolitan Patriots', in P. Cheah and B. Robbins (eds), *Cosmopolitics: Thinking and Feeling beyond the Nation*. Minneapolis, MN: University of Minnesota Press, pp. 91–114.

Appiah, K.A. (2005) *The Ethics of Identity*. Princeton, CT: Princeton University Press.
Appiah, K.A. (2006) *Cosmopolitanism. Ethics in a World of Strangers*. New York: W.W. Norton.
Appiah, K.A. (2007) *Cosmopolitanism: Ethics in a World of Strangers*, New York: W.W. Norton.
Archibugi, D. (1998) 'Principles of Cosmopolitan Democracy', in D. Archibugi, D. Held and M. Köhler (eds), Re-Imagining Political Community: Studies in Cosmopolitan Democracy. Cambridge, Polity Press, pp. 198–230.
Archibugi, D. (2003) 'Cosmopolitical Democracy', in D. Archibugi (ed.), *Debating Cosmopolitics*. London and New York: Verso, pp. 1–15.
Arendt, H. (1976) *The Origins of Totalitarianism*. New York: Harvest.
Ashis, N. (1998) *The Intimate Enemy: Loss and Recovery of Self under Colonialism*. Delhi: Oxford University Press.
Bahovec, T. (ed.) (2003) *Eliten und Nationwerdung: die Rolle der Eliten bei der Nationalisierung der Kärntner Slovenen*. Klagenfurt: Hermagoras.
Bajruši, R. (2007) 'Svi grijesi Zdenke Babić – Petričević', *Nacional*, 21/06.
Bakalian, A. (1994) *Armenian-Americans: From Being to Feeling Armenian*. New Brunswick: Transaction Publishers.
Baker, G. (2007) 'Commentaries: Debating (De)territorial Governance: Post-territorial Politics and the Politics of Difference', *Area*, 39(1): 109–12.
Bardenstein, C. (1999) 'Trees, Forests, and the Shaping of Palestinian and Israeli Collective Memory', in M. Bal, J. Crewe and L. Spitzer (eds), *Acts of Memory: Cultural Recall in the Present*. Hanover: Dartmouth College, pp. 148–68.
Barrington, L.W., E.S. Herron and B.D. Silver (2003) 'The Motherland is Calling. Views of Homeland among Russians in the Near Abroad', *World Politics*, 55(2): 290–313.
Barry, K. (2006) 'Symposium – Home and Away: The Construction of Citizenship in an Emigration Context', *University Law Review*, 81(1): 11–59.
Basch, L., N.G. Schiller and C.S. Blanc (1994) *Nations Unbound Transnational Projects. Postcolonial Predicaments, and Deterritorrialized Nation-States*. New York: Gordon and Breach.
Basu, P. (2005) 'Roots Tourism as Return Movement: Semantics and the Scottish Diaspora', in M. Harper (ed.), *Emigrant Homecomings. The Return Movement of Emigrants, 1600–2000*. Manchester: Manchester University Press, pp. 131–51.
Bauböck, R. (2005) 'Expansive Citizenship-Voting beyond Territory and Membership', *PS: Political Science & Politics*, 38(4): 683–87.
BBC (2001) *Born Abroad: An Immigration Map of Britain*. Online. Available at: http://www.news.bbc.co.uk/1/shared/spl/hi/uk/05/born_abroad/countries/html/romania.stm (accessed 20 January 2012).
Beck, U. (2002a) 'The Cosmopolitan Society and its Enemies', *Theory, Culture & Society*, 19(1–2): 17–44.
Beck, U. (2002b) 'The Terrorist Threat: World Risk Society Revisited', *Theory, Culture & Society*, 4(19): 39–55.
Beck, U. (2006) *The Cosmopolitan Vision*. Malden, MN and Cambridge: Polity Press.
Beck, U. and E. Grande (2007) 'Cosmopolitanism: Europe's Way Out of Crisis', *European Journal of Social Theory* 10(1): 67–85.
Beck, U. and E. Grande (2008) *Cosmopolitan Europe*. Cambridge: Polity Press.
Beck, U. and N. Sznaider (2006) 'Unpacking Cosmopolitanism for the Social Sciences: a Research Agenda', *The British Journal of Sociology*, 57(1): 1–23.
Benhabib, S. (2005) *Another Cosmopolitanism*. Oxford: Oxford University Press.

Benhabib, S. (2007) 'Twilight of Sovereignty or the Emergence of Cosmopolitan Norms? Rethinking Citizenship in Volatile Times', *Citizenship Studies*, 11(1): 19–36.

Berkan, W. (1924) *Życiorys własny*. Poznań: Fiszer i Majewski.

Bhabha, H.K. (1990) 'The Third Space', in J. Rutherford (ed.), *Identity: Community, Culture, Difference*. London: Lawrence & Wishart, pp. 207–21.

Bhimji, F. (2008) 'Cosmopolitan Belonging and Diaspora: Second-Generation British Muslim Women Travelling to South Asia', *Citizenship Studies*, 12(4): 413–27.

Bigo, D. and R.B.J. Walker (2007) 'International, Political, Sociology', *International Political Sociology*, 1(1): 1–5.

Bilger, V. and I. van Liempt (2009) 'Introduction' in V. Bilger and I. van Liempt (eds), *The Ethics of Migration Research Methodology: Dealing with Vulnerable Immigrants*. Brighton, Portland: Sussex Academic Press.

Bird, E.S. (2003) *The Audience in Everyday Life: Living in a Media World*. New York and London: Routledge.

Birman, D. (2006) 'Ethical Issues in Research with Immigrants and Refugees' in J.E. Trimble and C.B. Fisher (eds), *The Handbook of Ethical Research with Ethnocultural Populations and Communities*. Thousand Oaks, CA: Sage Publications.

Blitz, B.K. (2005) 'Refugee Returns, Civic Differentiation, and Minority Rights in Croatia 1991–2004', *Journal of Refugee Studies*, 18(3): 362–86.

Boehm, M.H. (1931) 'Cosmopolitanism', in E.R.A. Seligman (ed.), *Encyclopaedia of the Social Sciences*, Vol. 4. London: Macmillan.

Bolak, H. (1997) 'Studying one's own in the Middle East: Negotiating Gender and Self-Other Dynamics in the Field', in R. Hertz (ed.), *Reflexivity and Voice*. London: Sage Publications, pp. 95–118.

Borkert, M. and C. De Tona (2006) 'Stories of HERMES: a Qualitative Analysis of (Qualitative) Questions of Young Researchers in Migration and Ethnic Studies in Europe', *Forum Qualitative Sozialforschung/Forum: Qualitative Social Research*, 7(3): Art. 9. Online. Available at: http://www.qualitative-research.net/index.php/fqs/article/view/133/288 (accessed 15 October 2011).

Bosniak, L. (2000) 'Citizenship Denationalized', *Indiana Journal of Global Legal Studies*, 7(2): 447–509.

Brah, A. (1996) *Cartographies of the Diaspora*. London: Routledge.

Braverman, I. (2009) 'Planting the Promised Landscape: Zionism, Nature and Resistance in Israel/Palestine', *The Natural Resources Journal*, 49: 317–61.

Brubaker, R. (1996) *Nationalism Reframed: Nationhood and the National Question in the New Europe*. Cambridge: Cambridge University Press.

Brubaker, R. (2005) 'The "Diaspora" Diaspora', *Ethnic and Racial Studies*, 28(1): 1–19.

Brunner, O. (1982) *Geschichtliche Grundbegriffe: Historisches Lexikon zur Politisch-Sozialen Sprache in Deutschland,* Vol 3. Stuttgart: Klett-Cotta.

Burawoy, M. and K. Verdery (eds) (1999) *Uncertain Transition: Ethnographies of Change in the Postsocialist World*. New York and Oxford: Rowman & Littlefield Publishers.

Burchell, D. (1995) 'The Attributes of Citizens: Virtue, Manners and the Activity of Citizenship', *Economy and Society*, 24: 540–58.

Burrai, V. (2008). 'Kin-state Politics and Equal Treatment in Croatia', ASN Annual Convention, New York.

Busch, B. (2003) 'Shifting Political and Cultural Borders: Language and Identity in the Border Region of Austria and Slovenia', *European Studies: An Interdisciplinary Series in European Culture, History and Politics*, 19: 125–44.

Byford, A. (2009) '"The Last Soviet Generation" in Britain', in J. Fernandez (ed.), *Diasporas: Critical and Inter-Disciplinary Perspectives*. The Inter-Disciplinary Press. Online. Available at: http://www.inter-disciplinary.net/publishing/id-press/ebooks/diasporas/ (accessed 13 October 2009).

Calhoun, C. (2002) 'The Class Consciousness of Frequent Travellers: Toward a Critic of Actually Existing Cosmopolitanism', *South Atlantic Quarterly*, 101(4): 869–97.

Carter, S. (2005) 'The Geopolitics of Diaspora', *Area*, 37: 54–63.

Castles, S., and M.J. Miller (1993) *The Age of Migration. International Population Movements in the Modern World*. Basingstoke: Palgrave Macmillan.

Cauvet, P. (2011) 'Deterritorialisation, Reterritorialisation, Nations and States: Irish Nationalist Discourses on Nation and Territory before and after the Good Friday Agreement', *GeoJournal*, 76: 77–91.

Chandler, D. (2007) 'Commentaries: Debating (De)territorial Governance: The Possibilities of Post-territorial Political Community', *Area*, 39(1): 116–19.

Chandler, D. (2009) 'Critiquing Liberal Cosmopolitanism? The Limits of the Biopolitical Approach', *International Political Sociology*, 3(1): 53–70.

Cheah, P. (1998) 'Given Culture: Rethinking Cosmopolitical Freedom', in P. Cheah and B. Robbins (eds), *Transnationalism. Cosmopolitics: Thinking and Feeling beyond the Nation*. Minneapolis, MN: University of Minnesota Press, pp. 290–328.

Chernilo, D. (2006) 'Social Theory's Methodological Nationalism', *European Journal of Social Theory*, 9(1): 5–22.

Chinn, J., and R. Kaiser (1996) *Russians as the New Minority. Ethnicity and Nationalism in the Soviet Successor States*. Boulder, CO: Westview Press.

Clifford, J. (1994) 'Diasporas', *Cultural Anthropology*, 9(3): 302–38.

Clifford, J. (1997) *Routes: Travel and Translation in the Late Twentieth Century*. London: Harvard University Press.

Clifford, J. (1998) 'Mixed Feelings', in P. Cheah and B. Robbins (eds), *Cosmopolitics: Thinking and Feeling Beyond the Nation*, Minneapolis, MN: University of Minnesota Press, pp. 362–70.

Clifford, J. and G.E. Marcus (eds) (1986) *Writing Culture: The Poetics and Politics of Ethnography*. Berkeley, CA: University of California Press.

Cohen, C. (2004) *Planting Nature. Trees and the Manipulation of Environmental Stewardship in America*. Berkeley, CA: University of California Press.

Cohen, M. (1992) 'Rooted Cosmopolitanism', *Dissent*, 39(4): 478–83.

Cohen, R. (1996) 'Diasporas and the Nation-State: From Victims to Challengers', *International Affairs*, 72(3): 507–20.

Cohen, R. (1997) *Global Diasporas. An Introduction*. London: UCL Press.

Cohen, R. (2007) *Creolization and Diaspora – The Cultural Politics of Divergence and Some Convergence*. Reno, NV: University of Nevada Press.

Cohen, R. (2008) *Global Diasporas*, 2nd edn. London: Routledge.

Conradson, D. and A. Latham (2005a) 'Friendship, Networks and Transnationality in a World City: Antipodean Migrants in London', *Journal of Ethnic and Migration Studies*, 31(2): 287–305.

Conradson, D. and A. Latham (2005b) 'Transnational Urbanism: Attending to Everyday Practices and Mobilities', *Journal of Ethnic and Migration Studies*, 31(2): 227–33.

Cox, J.K. (2005) *Slovenia: Evolving Loyalties*. London: Routledge.

Crawford, B. (1996) 'Explaining Defection from International Cooperation: Germany's Unilateral Recognition of Croatia', *World Politics*, 48: 482–521.

Croatia, Republic of (1998 [1991]) *The Constitution of the Republic of Croatia*. Zagreb: Narodne novine.

Crotty, M. (1998) *The Foundations of Social Research: Meaning and Perspective in the Research Process*. London: Sage Publications.

Cummings, S.N. (2005) *Kazakhstan. Power and the Elite*. London: Tauris.

Cummings, S.N. (2006) 'Legitimation and Identification in Kazakhstan', *Nationalism and Ethnic Politics*, 12(2): 177–204.

Czebatul, M. (1999) *Opowieść Marty z Szułcików*. Nowy Tomyśl: Biblioteka Publiczna Miasta i Gminy.

Dalby, S. and G. Toal (1998) *Rethinking Geopolitics*. New York: Routledge.

Danielewicz-Kerski, D. and M. Górny (eds) (2008) *Berlin. Polnische Perspektiven. 19.–21. Jahrhundert*. Berlin: Berlin Story Buchhandlung & Verlag.

Darieva, T. (2006) 'Bringing the Soil back to the Homeland. Reconfigurations of Representation of Loss in Armenia', *Comparativ. Leipziger Beiträge zur Universalgeschichte und vergleichenden Gesellschaftsforschung. Heft 3, Transfer lokalisiert: Konzepte, Akteure, Kontexte*, 87–101.

Darieva, T. (2011) 'Rethinking Homecoming. Diasporic Cosmopolitanism in Post-socialist Armenia', *Ethnic and Racial Studies*, 34(3): 490–508.

Dave, B. (2007) *Kazakhstan. Ethnicity, Language and Power*. London: Routledge.

Davis, S. and S. Sabol (1998) 'The Importance of Being Ethnic: Minorities in Post-Soviet States – the Case of Russians in Kazakstan', *Nationalities Papers*, 26(3): 473–91.

Dickenson, J. and M.J. Andrucki *et al.* (2008) 'Introduction: Geographies of Everyday Citizenship', *ACME: An International E-Journal for Critical Geographies*, 7(2): 100–12.

Denzin, N. and Y. Lincoln (2005) 'Introduction. The Discipline and Practice of Qualitative Research', in N. Denzin and Y. Lincoln (eds), *The Sage Handbook of Qualitative Research*, 3rd edn, Thousand Oaks, CA: Sage Publications, pp. 1–32.

Diener, A.C. (2004) *Homeland Conceptions and Ethnic Integration among Kazakhstan's Germans and Koreans*. Lewiston, NY: E. Mellen Press.

Dietz, B. (2006) 'Aussiedler in Germany: From Smooth Adaptation to Tough Integration', in L. Lucassen, D. Feldman, and J. Oltmer (eds), *Immigrant Integration in Western Europe, Then and Now*. Amsterdam: Amsterdam University Press, pp. 116–36.

Dika, M.A., C. Helton, and J. Omejec (1998) 'The Citizenship Status of Citizens of the Former SFR Yugoslavia after its Dissolution', *Croatian Critical Law Review*, 3(1–2): 1–259.

Dikec, M. (2002) 'Pera Peras Poros – Longings for Spaces of Hospitality', *Theory, Culture & Society*, 19(1–2): 227–49.

Dikec, M., N. Clark and C. Barnett (2009) 'Extending Hospitality: Giving Space, Taking Time', *Paragraph*, 32(1): 1–14.

Diogenes, L. (1972, [*c*200 BC]) *Lives of Eminent Philosophers*, Vol. I, transl. R.D. Hicks. London: Heinemann.

Düvell, F., A. Triandafyllidou and B. Vollmer, B (2008) *Ethical Issues in Irregular Migration Research*. European Commission: Clandestino. Online. Available at: http://irregular-migration.net//typo3_upload/groups/31/4.Background_Information/4.1.Methodology/EthicalIssuesIrregularMigration_Clandestino_Report_Nov09.pdf (accessed 10 October 2011).

England, K. (1994) 'Getting Personal: Reflexivity, Positionality, and Feminist Research', *The Professional Geographer*, 46(1): 80–9.

Erdei, I. (2008) *Pacuraru: Doua milioane de romani sint angajati in strainatate*. Online. Available at: http://www.standard.ro/articles/print_article/49543 (Accessed 20 January 2012).

Esman, M.J. (2009) *Diasporas in the Contemporary World*. Cambridge: Polity Press.

Faist, T. (2004) 'Towards a Political Sociology of Transnationalization. The State of the Art in Migration Research', *European Journal of Sociology*, 45(3): 331–66.

Faist, T. (2007) 'Transnationalism and Development(s): Towards a North–South Perspective', *COMCAD Working Paper*, No. 16. Bielefeld: Centre on Migration, Citizenship and Development.

Falzon, M.A. (2003) '"Bombay, Our Cultural Heart": Rethinking the Relation between Homeland and Diaspora', *Ethnic and Racial Studies*, 26(4): 662–83.

Favell, A. (2008) *Eurostars and Eurocities: Free Movement and Mobility in an Integrating Europe*. Oxford: Blackwell.

Featherstone, M. (2002) 'Cosmopolis. An Introduction', *Theory, Culture & Society*, 19(1–2): 1–16.

Fierman, W. (1998) 'Language and Identity in Kazakhstan: Formulations in Policy Documents 1987–1997', *Communist and Post-Communist Studies*, 31(2): 171–86.

Fine, R. (2003) 'Taking the "ism" out of Cosmopolitanism', *European Journal of Social Theory*, 6(4): 451–70.

Fitzgerald, D. (2006) 'Rethinking Emigrant Citizenship', *New York University Law Review*, 81(1): 101–25.

Freeman, M. (1993) *Rewriting the Self: History, Memory, Narrative (Critical Psychology)*. London and New York: Routledge.

Friedman, J. (1995) 'Global System, Globalisation and the Parameters of Modernity', in M. Featherstone, S. Lash and R. Robertson (eds), *Global Modernities*. London: Sage Publications, pp. 69–90.

Gadamer, H.G. (1960) *Wahrheit und Methode: Grundlegung einer philosophischen Hermeneutik*. Tübingen: Mohr.

Gamlen, A. (2009) 'The Emigration State and the Modern Geopolitical Imagination', *Political Geography*, 27(8): 840–56.

Ganga, D. and S. Scott (2006) 'Cultural "Insiders" and the Issue of Positionality in Qualitative Migration Research: Moving "Across" and Moving "Along" Researcher–Participant Divides', *Forum: Qualitative Social Research*, 7(3). Online. Available at: http://www.qualitative-research.net/index.php/fqs/article/view/134/289 (accessed 24 July 2011).

Georgiou, M. (2007) 'Transnational Crossroads for Media and Diaspora', in O. Guedes-Bailey and R. Harindranath (eds), *Lives and the Media: Reimagining Diasporas*. Basingstoke: Palgrave.

Gerber, D.A. (2006) *Authors of their Lives. The Personal Correspondence of British Immigrants to North America in the Nineteenth Century*. New York and London: New York University Press.

Gerhardt, V. (1995) *Immanuel Kant's Entwurf 'Zum ewigen Frieden': Eine Theorie der Politik*. Darmstadt: Wissenschaftliche Buchgesellschaft.

Gilroy, P. (1987) *There Ain't No Black in the Union Jack: The Cultural Politics of Race and Nation*. London: Routledge.

Gilroy, P. (1994) 'Diaspora', *Paragraph*, 17(1): 207–12.

Glick Schiller, N. (2005) 'Blood and Belonging: Long-distance Nationalism and the World Beyond', in S. McKinnon and S. Silverman (eds), *Complexities: Beyond Nature and Nurture*. Chicago, IL: University of Chicago Press, pp. 448–67.

Glick Schiller, N. (2008) 'Beyond Methodological Ethnicity: Local and Transnational Pathways of Immigrant Incorporation', *Willy Brandt Series of Working Papers in International Migration and Ethnic Relations*, No 2, Online. Available at: http://muep.mah.se/bitstream/handle/2043/7491/WB%202_08%20MUEP.pdf?sequence=3(accessed 23 October 2011).

Glick Schiller, N. and G. Fouron, G. (1999) 'Terrains of Blood and Nation: Haitian Transnational Social Fields', *Ethnic and Racial Studies*, 22(2): 340–66.

Glick Schiller, N. and G. Fouron (2001) *Georges Woke up Laughing: Long Distance Nationalism and the Search for Home*. Durham, NC: Duke University Press.

Glick Schiller, N. and P. Levitt (2004) 'Conceptualizing Simultaneity: A Transnational Social Field Perspective on Society', *International Migration Review*, 38(3): 1002–39.

Glick Schiller, N., A. Caglar and T. Guldbrandsen (2006) 'Beyond the Ethnic Lens: Locality, Globality, and Born-again Incorporation', *American Ethnologist*, 33(4): 612–33.

Glick Schiller, N., T. Darieva and S. Gruner-Domic (2011) 'Defining Cosmopolitan Sociability in a Transnational Age. An Introduction', *Ethnic and Racial Studies*, 34(3): 399–418.

Gotfryd, A. (2005) *Der Himmel in den Pfützen: Ein Leben zwischen Galizien und dem Kurfürstendamm*. Berlin: WJS Verlag.

Grahl-Madsen, A. (1983) 'The League of Nations and the Refugees', *The League of Nations in Retrospect*. Berlin: Walter de Gruyter, pp. 358–68.

Grant, B. (2010) '"Cosmopolitan Baku"', *Ethnos*, 75(2): 123–47.

Grant, T.D. (1999) *The Recognition of States: Law and Practice in Debate and Evolution*. Westport, CT: Praeger Publishers.

Gray, B. (2006) 'Redefining the Nation through Economic Growth and Migration: Changing Rationalities of Governance in the Republic of Ireland?', *Mobilities*, 1(3): 353–72.

Guillemin, M. and L. Gillam, (2004) 'Ethics, Reflexivity, and "Ethically Important Moments" in Research', *Qualitative Inquiry*, 10(2): 261–80.

Habermas, J. and M. Pensky (2001) *The Postnational Constellation Political Essays, Studies in Contemporary German Social Thought*. London: MIT Press.

Halbwachs, M. (1992) *On Collective Memory*, translated and edited by L.A. Coser. Chicago, IL: University of Chicago Press.

Halemba, A. (2011) 'National, Transnational or Cosmopolitan Heroine? The Virgin Mary's Apparitions in Cotemporary Europe', *Ethnic and Racial Studies*, 34(3): 454–70.

Hall, S. (1990) 'Cultural Identity and Diaspora', in J. Rutherford (ed.), *Identity: Community, Culture, Difference*. London: Lawrence and Wishart, pp. 222–37.

Hall, S. (1996) 'New Ethnicities', in D. Morley and K. H. Chen (eds), *Stuart Hall: Critical Dialogues in Cultural Studies*,. London: Routledge, pp. 441–49.

Hannerz, U. (1990) 'Cosmopolitans and Locals in World Culture', in M. Featherstone (ed.), *Global Culture. Nationalism. Globalization and Modernity*. London: Sage Publications, pp. 237–52.

Hansen, R. (2009) 'The Poverty of Postnationalism: Citizenship, Immigration, and the New Europe', *Theory & Society*, 38(1): 1–24.

Harvey, J. (2006) 'Return Dynamics in Bosnia and Croatia: A Comparative Analysis', *International Migration*, 44(3): 89–144.

Helm, T. (2008) 'East European Workers Quit UK to Head Home', *The Observer*, 21 September 2008. Online. Available at: http://www.guardian.co.uk/world/2008/sep/21/poland.nhs (accessed 15 September 2009).

Hindess, B. (2001) 'Citizenship in the International Management of Populations', in D. Meredyth and J. Minson (eds), *Citizenship and Cultural Policy*. London: Sage Publications, pp. 92–103.

Hirsch, F. (2005) *Empire of Nations. Ethnographic Knowledge and the Making of the Soviet Union*. Ithaca, NY: Cornell University Press.

Hockenos, P. (2003) *Homeland Calling: Exile Patriotism and the Balkan Wars*. Ithaca, NY: Cornell University Press.

Hollinger, D.A. (1995) *Postethnic America: Beyond Multiculturalism*. New York: Basic Books.

Hollinger, D.A. (2001) 'Not Universalists, Not Pluralists: The New Cosmopolitans Find Their Own Way', *Constellations*, 8(2): 236–48.

Holm-Hansen, J. (1999) 'Political Integration in Kazakstan'. in P. Kolstø (ed.), *Nation-Building and Ethnic Integration in Post-Soviet Societies: An Investigation of Latvia and Kazakstan*. Boulder, CO: Westview Press, pp. 153–226.

Holsey, B. (2004) 'Transatlantic Dreaming: Slavery, Tourism and Diasporic Encounters', in F. Markowitz and A. Stefansson (eds), *Homecomings. Unsettling Paths of Return*. Lanham, MD: Lexington Books, pp. 166–82.

Human Rights Watch (1995) *Civil and Political Rights in Croatia*. New York: Human Rights Watch.

Human Rights Watch (2003) *Broken Promises: Impediments to Refugee Return to Croatia*. New York: Human Rights Watch.

Human Rights Watch (2004) *ICTY Justice at Risk: War Crimes Trials in Croatia, Bosnia And Herzegovina, and Serbia and Montenegro*. New York: Human Rights Watch.

Human Rights Watch (2006) *Croatia: A Decade of Disappointment. Continuing Obstacles to the Reintegration of Serb Returnees*, Vol. 18, No. 7(D).

Humphrey, C. (2004) 'Cosmopolitanism and Kosmopolitizm in the Political Life of Soviet Citizens', *Focaal, European Journal of Anthropology*, 44: 138–52.

Huntford, R. (1997) *Nansen: The Explorer as Hero*. London: Duckworth.

Hüwelmeier, G. (2011) 'Socialist Cosmopolitanism Meets Global Pentecostalism: Charismatic Christianity among Vietnamese Migrants after the Fall of the Berlin Wall', *Ethnic and Racial Studies*, 34(3):436–53.

Imeri, Sh. (2006) *Rule of Law in the Countries of the Former SFR Yugoslavia and Albania: Between Theory and Practice*. Gostivar: Association for Democratic Initiatives.

International Crisis Group (2002) *A Half-Hearted Welcome: Refugee Returns to Croatia*, Balkans Report No. 138. Zagreb/Brussels: International Crisis Group.

Iosifides, T. (2011) *Qualitative Methods in Migration Studies: A Critical Realist Perspective*. London: Ashgate.

Ishkanyan, A. (2008) *Democracy Building and Civil Society in Post-Soviet Armenia*. New York: Routledge.

Itzighson, J. (2000) 'Immigration and the Boundaries of Citizenship: The Institutions of Immigrants' Political Transnationalism', *The International Migration Review: IMR*, 34(4): 1126–54.

'Izbori 2005 – Arhiva'. Online. Available at: http://www.izbori.hr/2005Pred/ (accessed 14 May 2011).

'Izbori 2007 – Rezultati', Online. Available at: http://www.izbori.hr/izbori/izbori07.nsf/FI?OpenForm (accessed 14 May 2011).

Jacobson, D. (1996) *Rights across Borders: Immigration and the Decline of Citizenship*. Baltimore, MD: Johns Hopkins University Press.

Jacobsen, K. and L. Landau (2003) 'Researching Refugees: Some Methodological and Ethical Considerations in Social Science and Forced Migration', *New Issues in Refugee Research, UNCHR Working Paper*, No. 90, June.

Jinga, I. (2012) Speech given at the opening of the Romanian House in the Olympics Village, 29 July 2012, London. Available at: http://www. facebook.com/photo. php?v=1015080190028809. Accessed on 8 August 2012.

Joppke, C. (2003) 'Citizenship between De and Re-Ethnicization', *European Journal of Sociology*, 44(3): 429–58.

Jubulis, M.A. (2001) *Nationalism and Democratic Transition. The Politics of Citizenship and Language in Post-Soviet Latvia*. Lanham, MD: University Press of America.

Kaldor, M. (1996) 'Cosmopolitanism versus Nationalism: The New Divide?' in R. Caplan and J. Feffer (eds), *Europe's New Nationalism*. Oxford: Oxford University Press, pp. 42–58.

Kant, I. (1970 [1795]) 'Perpetual Peace: A Philosophical Sketch', in H. Reiss (ed.), *Kant's Political Writings*. Cambridge: Cambridge University Press, pp. 93–130.

Kant, I. (1992 [1795]) *Zum Ewigen Frieden. Ein Philosophisches Entwurf.* Hamburg: Mainer.

Kant, I. (2005) 'Towards Perpetual Peace', in P. Kleingeld *et al.* (eds), *Towards Perpetual Peace and Other Writings on Politics, Peace, and History*. New York: Yale University Press.

Kanter, R.M. (1995) *World Class: Thriving Locally in the Global Economy*. New York: Simon and Schuster.

Kasapović, M. (1996) '1995 Parliamentary Elections in Croatia', *Electoral Studies*, 15(2): 269–74.

Kasapović, M. (1999) 'Znaju li uopće koliko glasova košta jedan mandat?', *Globus*, 14/05: 10–11.

Kastoryano, R. (2007) 'Transnational Nationalism: Redefining Nation and Territory', in S. Benhabib, I. Shapiro and D. Petranovic (eds), *Identities, Affiliations, and Allegiances*. Cambridge: Cambridge University Press, pp. 159–80.

Kay, R. and M. Kostenko (2008) 'Men in Crisis or in Critical Need for Support? Insights from Russia and the UK', in M. Flynn, R. Kay and J. Oldfield (eds), *Trans-National Issues, Local Concerns and Meanings of Post-Socialism: Insights from Russia, Central Eastern Europe, and Beyond*. New York and Plymouth: University Press of America, pp. 101–24.

Keith, M. (1992) 'Angry Writing: (Re)presenting the Unethical World of the Ethnographer', *Environment and Planning D: Society and Space*, 10(5): 551–68.

Kerski, B. (2008) 'Homer auf dem Potsdamer Platz. Ein Berliner Essay' in D. Danielewicz-Kerski and M. Górny (eds), *Berlin. Polnische Perspektiven 19. – 21. Jahrhundert*. Berlin: Berlin Story Buchhandlung & Verlag.

Kharkhordin, O. (1999) *The Collective and the Individual in Russia: A Study of Practices*. Berkeley, CA: University of California Press.

Kharkhordin, O. (ed.) (2009) *Druzhba: ocherki po teorii praktik* [Friendship: Essays on the Theory of Practices]. St Petersburg: European University at St Petersburg Press.

Kimmerling, B. (1983) *Zionism and Territory: The Socio-territorial Dimensions of Zionist Politics*,. Berkeley, CA: Institute of International Studies University of California.

King, R. and Christou, A. (2008) 'Cultural Geographies of Counter-Diasporic Migration: The Second Generation Returns "Home"', *Sussex Migration Working Paper*, No. 45.

King, R. and Christou, A. (2010) 'Cultural geographies of counter-diasporic migration: perspectives from the study of second-generation 'returnees' to Greece', *Population, Space and Place*, 16(2): 103–19.

Kleingeld, P. (1999) 'Six Varieties of Cosmopolitanism in Late Eighteenth-Century Germany', *Journal of the History of Ideas*, 60(3): 505–24.

Klemenčič, M. (2009) 'The International Community and the FRY/Belligerents, 1989–1997', in C. Ingrao and T.A. Emmert (eds), *Confronting the Yugoslav controversies: A Scholars' Initiative*. West Lafayette, IN: Purdue University Press, pp. 153–98.

Kofman, E. (2005) 'Figures of the Cosmopolitan: Privileged Nationals and National Outsiders', *Innovation*, 18(1): 83–97.

Köhler, M. (1998) 'From the National to the Cosmopolitan Public Sphere', in D. Archibugi, D. Held and M. Köhler (eds), *Re-imagining Political Community: Studies in Cosmopolitan Democracy*. Oxford: Polity Press, pp. 231–5.

Kopnina, H. (2005) *East to West Migration: Russian Migrants in Western Europe*. Aldershot: Ashgate.

Korobkov, A.V. (2008) 'Post-Soviet Migration: New Trends at the Beginning of the Twenty-First Century', in C.J. Buckley, B.A. Ruble and E.T. Hofman (eds), *Migration, Homeland and Belonging in Eurasia*. Washington, DC: Woodrow Wilson Center Press, pp. 69–98.

Kosin, M. (1998) 'Slovenska manjšina v slovensko-italijanskih odnosih', *Razprave in gradivo*, 33: 57–97.

Kosmarskaya, N. (2006) *"Deti imperii" v post-sovetskoi Tsentralnoi Azii*. Moscow: Natalis.

Krynicki, R. (2008–2009) 'Eine Insel der Freiheit', *Dialog. Deutsch–Polnisches Magazin*, No. 85–86.

Kunz, R. (2011) *The Political Economy of Global Remittances Gender and Governmentality*, RIPE Series in Global Political Economy. New York: Routledge.

Laitin, D.D. (1998) *Identity in Formation. The Russian-Speaking Populations in the Near Abroad*. Ithaca, NY: Cornell University Press.

Larner, W. (2007) 'Expatriate Experts and Globalising Governmentalities: The New Zealand Diaspora Strategy', *Transactions of the Institute of British Geographers*, 32(3): 331–45.

Latour, B. (2004) 'Whose Cosmos, Which Cosmopolitics? Comments on the Peace Terms of Ulrich Beck', *Common Knowledge*, 10(3): 450–62.

Lettevall, R. (2008) 'The Idea of Kosmopolis: Two Kinds of Cosmopolitanism', in R. Lettevall and M.K. Linder (eds), *The Idea of Kosmopolis: History, Philosophy and Politics of World Citizenship*. Stockholm: Södertörn Academic Studies, pp. 13–30.

Lettevall, R. (2011) 'On the Historicity of Concepts: The Examples of Patriotism and Cosmopolitanism in Ellen Key', in H. Ruin and A. Ers (eds), *Rethinking Time: Essays on History, Memory and Representation*. Stockholm: Södertörn Philosophical Studies 9, pp. 179–88.

Lettevall, R. (forthcoming) 'Neutrality and Humanitarianism: Fridtjof Nansen and the Nansen Passports', in R. Lettevall, G. Somsen, and S. Widmalm (eds), *Neutrality in Twentieth-Century Europe: Intersections of Science, Culture, and Politics after the First World War*. London: Routledge.

Levin, N. (1990) *The Jews in the Soviet Union since 1917: paradox of survival*. New York: New York University Press.

Levitt, P. (1998) 'Social Remittances: Migration Driven Local-Level Forms of Cultural Diffusion', *International Migration Review*, 32(4): 926–48.

Levitt, P. (2001) *The Transnational Villagers.* Berkeley, CA: University of California Press.

Levitt, P. and R. de la Dehesa (2003) 'Transnational Migration and the Redefinition of the State: Variations and Explanations', *Ethnic and Racial Studies*, 26(4): 587–611.

Levitt, P. and M. Waters (2002) *The Changing Face of Home. The Transnational Lives of the Second Generation.* New York: Sage Publications.

Linklater, A. (1998) 'Cosmopolitan Citizenship', *Citizenship Studies*, 2(1): 23–41.

McDowell, L. (1992) 'Doing Gender: Feminism, Feminists and Research Methods in Human Geography', *Transactions of the Institute of British Geographers, New Series*, 17(4): 399–416.

McDowell, L. (2008) 'On the Significance of Being White: European Migrant Workers in the British Economy in the 1940s and 2000s', in C. Dwyer and C. Bressey (eds), *New Geographies of Race and Racism.* Aldershot: Ashgate: pp. 51–67.

Malia, M.E. (1994) *The Soviet Tragedy: History of Socialism in Russia 1917–1991.* New York: Free Press.

Malkki, L. (1997) 'National Geographic: The Rooting of Peoples and the Territorialization of National Identity among Scholars and Refugees', in A. Gupta and J. Ferguson (eds), *Culture, Power, Place: Explorations in Critical Anthropology.* Durham, NC: Duke University Press, pp. 52–74.

Mamattah, S. (2008). 'Ethnic German Community in a Russian City: The Local, the Global and Identity Formation', in M. Flynn, R. Kay and J. Oldfield (eds), *Trans-National Issues, Local Concerns and Meanings of Post-Socialism: Insights from Russia, Central Europe, and Beyond.* Lanham, MD: University of America Press.

Mandel, R. (2002) 'Seeding Civil Society', in C. M. Hann (ed.), *Postsocialism: Ideals, Ideologies and Practices in Eurasia.* London and New York: Routledge, pp. 279–96.

Mandel, R. (2008) 'Germans, Jews, or Russians? Diaspora and the Post-Soviet Transnational Experience', in C.J. Buckley, B.A. Ruble and E.T. Hofman (eds), *Migration, Homeland and Belonging in Eurasia.* Washington, DC: Woodrow Wilson Center Press, pp. 303–26.

Marrus, M.R. (1985) *The Unwanted: European Refugees from the First World War through the Cold War.* Philadelphia, PA: Temple University Press.

Marsden, M. (2008) 'Muslim Cosmopolitans? Transnational Life in Northern Pakistan', *The Journal of Asian Studies*, 67(1): 213–47.

Martin, T.D. (1998) 'The Origins of Soviet Ethnic Cleansing', *The Journal of Modern History*, 70(4): 813–61.

Martin, T.D. (2001) *The Affirmative Action Empire. Nations and Nationalism in the Soviet Union, 1923–1939.* Ithaca, NY: Cornell University Press.

Martins, H. (1974) 'Time and Theory in Sociology', in J. Rex (ed.), *Approaches to Sociology: An Introduction to Major Trends in British Sociology.* London: Routledge & Kegan Paul, pp. 246–94.

Mecanović (1999) 'Izbori i Dijaspora', *Pravni Vjesnik*, 14(1–2): 7–18.

Meier, V. (1999) *Yugoslavia: A History of Its Demise.* London and New York: Routledge.

Melkonyan, E. (2010) *Armyanskiy vseobshiy blagotvoritelnyi soyuz. Neokonchennaya istoriya.* Yerevan: Tigran Mets.

Merton, R. (1972) 'Insiders and Outsiders: A Chapter in the Sociology of Knowledge', *American Journal of Sociology*, 78(1): 9–47.

Merton, R.K. (1964 [1949]) *Social Theory and Social Structure.* London: The Free Press.

Miller, D. (2008) 'London: Nowhere in Particular'. Online. Available at: http://www.ucl.ac.uk/anthropology/staff/d_miller/mil-17 (accessed 7 September 2010).

Minahan, J. (1998) *Miniature Empires: A Historical Dictionary of the Newly Independent States*. Westport, CT: Greenwood Press.

Molz, J.G. (2005) 'Getting a Flexible Eye: Round the World Travel and Scales of Cosmopolitan Citizenship', *Citizenship Studies*, 9(5): 517–31.

Morawska, E. (2008) 'National Identities of Polish (Im)Migrants in Berlin, Germany: Four Varieties, Their Correlates and Implications', in W. Spohn, A. Triandafyllidou (eds), *Europeanisation, National Identities and Migration: Changes in Boundary Constructions between Western and Eastern Europe*. London: Routledge, pp. 171–91.

Mullings, B. (1999) 'Insider or Outsider, Both or Neither: Some Dilemmas of Interviewing in a Cross-cultural Setting', *Geoforum* 30: 337–50.

Nieswand, B. (2009) 'Development and Diaspora: Ghana and its Migrants', *Sociologus. Zeitschrift fuer empirische Ethnosoziologie und Ethnopsychologie*, 59(1): 17–31.

Niewrzęda, K. (2005) *Czas przeprowadzki*. Szczecin: Forma Autorska.

Notar, B.E. (2008) 'Producing Cosmopolitanism at the Borderlands: Lonely Planeteers and "Local" Cosmopolitans in Southwest China', *Anthropological Quarterly*, 81(3): 615–50.

Novick, P. (2001) *The Holocaust and Collective Memory. The American Experience*. London: Bloomsbury.

Nowicka, E. (ed) (2011) *Blaski i cienie imigracji: problemy cudzoziemców w Polsce*. Warszawa: Wydawnictwa Uniwersytetu Warszawskiego.

Nowicka, E. and S. Łodziński (1990) *U progu otwartego świata: poczucie polskości i nastawienia Polaków wobec cudzoziemców w latach 1988–1998*. Kraków: Nomos.

Nowicka M. and M. Rovisco (eds) (2009) *Cosmopolitanism in Practice*. Aldershot: Ashgate.

Nussbaum, M. (1996) 'Patriotism and Cosmopolitanism', in J. Cohen (ed.), *For Love of Country: Debating the Limits of Patriotism*. Chicago, IL: University of Chicago Press, pp. 3–20.

Oakley, A. (1981) 'Interviewing Women: A Contradiction in Terms', in H. Robertson (ed.), *Doing Feminist Research*. London: Routledge and Kegan Paul, pp. 31–61.

Office for National Statistics (2011) *Population by Country of Birth and Nationality April 2010 to March 2011*. Online. Available at: http://www.ons.gov.uk/ons/publications/re-reference-tables.html?edition=tcm%3A77-235204 (accessed 20 January 2012).

Oka, N. (2006) 'The "Triadic Nexus" in Kazakhstan: A Comparative Study of Russians, Uighurs, and Koreans', in O. Leda (ed.), *Beyond Sovereignty: From Status Law to Transnational Citizenship?* Sapporo: Hokkaido University, Slavic Research Centre, pp. 359–380.

Olwig, K. (2004) '"Place, Movement and Identity": Processes of Inclusion and Exclusion in a Caribbean Family', in W. Kokot (ed.), *Diaspora, Identity and Religion*. London: Routledge, pp. 53–71.

Omejec, J. (1998) 'Initial Citizenry of the Republic of Croatia at the Time of the Dissolution of Legal Ties with the SFRY, and Acquisition and Termination of Croatian Citizenship', *Croatian Critical Law Review*, 3(1–2): 99–127.

Ong, A. (1998) 'Flexible Citizenship among Chinese Cosmopolitans', in P. Cheah and B. Robbins (eds), *Cosmopolitics. Thinking and Feeling beyond the Nation*. Minneapolis, MN: University of Minnesota Press, pp. 134–62.

Ong, A. (2000) *Flexible Citizenship: The Cultural Logics of Transnationality*. Durham, NC: Duke University Press.

OSCE (1997) 'Statement, Presidential Election in the Republic of Croatia, 15 June 1997'. Online. Available at: http://www.osce.org/item/1536.html (accessed 23 March 2011).

OSCE (2000) 'Election of Representatives to the State Parliament, 2–3 January 2000', *International Election Observation Mission Preliminary Statement*. Online. Available at: http://www.osce.org/odihr/elections/croatia/15668online (accessed 23 March 2011).

OSCE/ODHIR (2000) 'Republic of Croatia. Extraordinary Presidential Elections 24 January and 7 February 2000', *OSCE/ODHIR Election Observation Mission Final Report*. Online. Available at: http://www.osce.org/odihr/elections/croatia/15670 (accessed 23 March 2011).

Panossian, R. (2006) *The Armenians. From Kings and Priests to Merchants and Commissars*. New York: Columbia University Press.

Pearce, K.E. (2011) 'Poverty in the South Caucasus', *Caucasus Analytical Digest*, 34: 2–12.

Petrosyan, H. (2001) 'The World as Garden', in L. Abrahamian and N. Sweezy (eds), *Armenian Folk Arts, Culture and Identity*. Bloomington, IN: Indiana University Press, pp. 25–32.

Peyrouse, S. (2007) 'Nationhood and the Minority Question in Central Asia. The Russians in Kazakhstan', *Europe-Asia Studies*, 59(3): 481–501.

Peyrouse, S. (2008) 'The "Imperial Minority": An Interpretative Framework of the Russians in Kazakhstan in the 1990s', *Nationalities Papers*, 36(1): 105–23.

Phillips, J. (1989) *Symbol, Myth, and Rhetoric: The Politics of Culture in an Armenian-American Population*. New York: AMS Press.

Pichler, F. (2009) '"Down-to-Earth" Cosmopolitanism: Subjective and Objective Measurements of Cosmopolitanism in Survey Research', *Current Sociology*, 57(5): 704–32.

Pichler, E., and O. Schmidtke (2004) 'Migranten im Spiegel des deutschen Mediendiskurses: "Bereicherung" oder "Belastung"?' in K. Eder, V. Rauer, and O. Schmidtke (eds), *Die Einhegung des Anderen: Türkische, polnische und russlanddeutsche Einwanderer in Deutschland*. Wiesbaden: Verlag für Sozialwissenschaften, pp. 49–76.

Pilkington, H. and A. Popov (2008) 'Cultural Production and the Transmission of Ethnic Tolerance and Prejudice: Introduction', *Anthropology of East Europe Review*, 26(1): 7–21.

Piore, M. (1980) *Birds of Passage: Migrant Labor and Industrial Societies*. New York: Cambridge University Press.

Piot, C. (1999) *Remotely Global*. Chicago, University of Chicago Press.

Pollock, S., H.K. Bhabha *et al.* (2000) 'Cosmopolitanisms', *Public Culture*, 12(3): 577–89.

Poppe, E. and L. Hagendoorn (2001) 'Types of Identification among Russians in the "Near Abroad"', *Europe–Asia Studies*, 53(1): 57–71.

Praszałowicz, D. (2006) 'Polish Berlin. Differences and Similarities with Poles in the Ruhr Area, 1860–1920', in L. Lucassen, D. Feldman, J. Oltmer (eds), *Paths of Integration: Migrants in Western Europe (1880–2004)*. Amsterdam: Amsterdam University Press, pp. 139–57.

Praszałowicz, D. (2010) *Polacy w Berlinie. Strumienie migracyjne i społeczności imigracyjne. Przegląd badań*. Kraków: Księgarnia Akademicka.

Praszałowicz, D. (2011) 'Poland', in K. Bade, P.C. Emmer, L. Lucassen, J. Oltmer (eds), *Encyclopedia of Migrations and Minorities in Europe from the 17th Century to the Present*. Cambridge: Cambridge University Press, pp.143–57.

Pries, L. (2001) 'The Disruption of Social and Geographic Space. Mexican–US Migration and the Emergence of Transnational Social Spaces', *International Sociology*, 16(1): 55–74.

Pugh, J., C. Hewett and D. Chandler (2007) 'Commentaries: Debating (De)territorial Governance: Debating (De)territorial Governance', *Area*, 39(1): 107–9.

Ragazzi, F. (2009a) 'The Croatian ''Diaspora Politics'' of the 1990s: Nationalism Unbound?', in U. Brunnbauer (ed.), *Transnational Societies, Transterritorial Politics. Migrations in the (Post-)Yugoslav Area. 19th–21st Centuries*. Munich: Oldenourg Verlag, pp. 145–68.

Ragazzi, F. (2009b) 'Governing Diasporas', *International Political Sociology*, 3(4), 378–97.

Ragazzi, F. and I. Štiks (2009) 'Croatian Citizenship: From Ethnic Engineering to Inclusiveness', in R. Baubock, W. Sievers and B. Perchinig (eds), *Citizenship and Migration in the New Member and Accession States of the E.U.* Amsterdam: University of Amsterdam Press, pp. 339–63.

Rakowski, K. (1901) *Kolonia polska w Berlinie*. Warszawa: Biblioteka Warszawska.

Rapport, N. and R. Stade (2010) 'A Cosmopolitan Turn or Return?', *Social Anthropology*, 15(2): 223–35.

Richardson, T. (2006) 'Living Cosmopolitanism? Tolerance, Religion and Local Identity in Odessa', in C. Hann (ed.), *The Postsocialist Religious Question: Faith and Power in Central Asia and East-Central Europe*. Halle: Lit Verlag, pp. 213–40.

Rival, L. (1998) 'Trees, from Symbols of Life and Regeneration to Political Artefacts', in L. Rival (ed.), *The Social Life of Trees. Anthropological Perspectives on Tree Symbolism*. Oxford: Berg, pp. 1–36.

Robertson, J. (2002) 'Reflexivity Redux: A Pithy Polemic on "Positionality"', *Anthropological Quarterly*, 75(4): 785–92.

Robbins, B. (1998a) 'Actually Existing Cosmopolitanism', in P. Cheah and B. Robbins (eds), *Cosmopolitics, Thinking and Feeling beyond the Nation*. Minneapolis, MN: University of Minnesota, pp. 1–19.

Robbins, B. (1998b) 'Comparative Cosmopolitanisms', in P. Cheah and B. Robbins (eds), *Cosmopolitics, Thinking and Feeling beyond the Nation*. Minneapolis, MN: University of Minnesota, pp. 246–64.

Robinson, K. (2008) 'Islamic Cosmopolitics, Human Rights and Anti-Violence Strategies in Indonesia', in P. Werbner (ed.), *Anthropology and the New Cosmopolitanism*. Oxford: Berg, pp. 111–34.

Rodgers, G. (2004) '''Hanging out" with Forced Migrants: Methodological and Ethical Issues', *Forced Migration Review*, 21: 48–49.

Römhild, R. (1998) *Die Macht des Ethnischen: Grenzfall Rußlanddeutsche. Perspektiven einer politischen Anthropologie*. Frankfurt: Lang.

Römhild, R. (2007) 'Fremdzuschreibungen – Selbstpositionierungen. Die Praxis der Ethnisierung im Alltag der Einwanderungsgesellschaft', in B. Schmidt-Lauber (ed.), *Ethnizität und Migration: Einführung in Wissenschaft und Arbeitsfelder*. Berlin: Reimer, pp. 157–77.

Rose, G. (1997) 'Situating Knowledges: Positionality, Reflexivities and Other Tactics', *Progress in Human Geography*, 21(3): 305–20.

Rose, K. (1932) *Wspomnienia berlińskie*. Warszawa: F. Hoesick.

Rosenthal, G. (1993) 'Reconstruction of Life Stories. Principles of Selection in Generating Stories for Narrative Biographical Interviews', *Narrative Study of Life*, 1: 59–91.

Rubin, D.C. (1995) 'Introduction', in D.C. Rubin (ed.), *Remembering Our Past: Studies in Autobiographical Memory*. Cambridge: Cambridge University Press.

Ruggie, J.G. (1993) 'Territoriality and Beyond: Problematizing Modernity in International Relations', *International Organization*, 47(1): 139–74.

Rumford, C. (2005) 'Cosmopolitanism and Europe: Towards a New EU Studies Agenda?', *Innovation*, 18(1): 1–9.

Rumford, C. (ed) (2007) *Cosmopolitanism and Europe*. Liverpool: Liverpool University Press.

Safran, W. (1991) 'Diasporas in Modern Societies: Myth of Homeland and Return', *Diaspora*, 1(1): 83–99.

Salay, C., K. Duich and International Foundation for Electoral Systems (1996) *Republic of Croatia: 1995 Election Observation Report*. Washington, DC: International Foundation for Election Systems.

Sassen, S. (2006) *Territory, Authority, Rights: From Medieval to Global Assemblages*. Princeton, NJ: Princeton University Press.

Schatz, E. (2000) 'The Politics of Multiple Identities: Lineage and Ethnicity in Kazakhstan', *Europe-Asia Studies* 52(3): 489–506.

Schleiermacher, F.D.E. (1984 [1800/1810]) 'Versuch einer Theorie des geselligen Betragens', in J. Rachold (ed.). *Schleiermacher. Philosophische Schriften*. Berlin: Union Verlag Berlin, pp. 39–64.

Scott, S. (2006) 'The Social Morphology of Skilled Migration: The Case of British Middle Class in Paris', *Journal of Ethnic and Migration Studies*, 32(7): 1105–129.

Sheffer, G. (2003) *Diaspora Politics: at Home Abroad*. Cambridge: Cambridge University Press.

Shlapentokh, V. (1989) *Public and Private Life of the Soviet People: Changing Values in post-Stalin Russia*. Oxford: Oxford University Press.

Shukla, S. (1997) 'Building Diaspora and Nation: The 1991 "Cultural Festival of India"', *Cultural Studies*, 2(2): 296–315.

Skran, C.M. (1995) *Refugees in Inter-war Europe: The Emergence of a Regime*. Oxford: Clarendon Press.

Skrbiš, Z. (2002) 'The Emotional Historiography of Venetologists: Slovene Diaspora, Memory, and Nationalism', *European Journal of Anthropology*, 39: 41–55.

Skrbiš, Z. (2007a) '"The First Europeans" Fantasy of Slovenian Venetologists: Emotions and Nationalist Imaginings', in M. Svašek (ed.), *Postsocialism: Politics and Emotions in Central and Eastern Europe*. Oxford: Berghahn Books, pp. 138–58.

Skrbiš, Z. (2007b) 'The Mobilized Croatian Diaspora: Its Role in Homeland Politics and War', in H. Smith and P. Stares (eds), *Diasporas in Conflict: Peace-makers or Peace-wreckers?* Tokyo, New York and Paris: United Nations University Press, pp. 218–38.

Skrbiš, Z., G. Kendala, and I. Woodward (2004) 'Locating Cosmopolitanism: Between Humanist Ideal and Grounded Social Category', *Theory, Culture & Society*, 21(6): 115–36.

Slezkine, Y. (1994) 'The USSR as a Communal Apartment, or How a Socialist State Promoted Ethnic Particularism', *Slavic Review*, 53(2): 414–52.

Smith, A.D. (1979) *Nationalism in the Twentieth Century*. Oxford: Martin Robertson.

Smith, A.D. (1986) *The Ethnic Origin of Nations*. New York: Basil Blackwell.

Smith, R.C. (2003a) 'Diasporic Memberships in Historical Perspective: Comparative Insights from the Mexican, Italian and Polish Cases', *International Migration Review*, 37(3): 724–59.

Smith, R.C. (2003b) 'Migrant Membership as an Instituted Process: Transnationalization, the State and the Extra-Territorial Conduct of Mexican Politics', *International Migration Review*, 37(2): 297–343.

Soysal, Y.N. (1994) *Limits of Citizenship: Migrants and Postnational Membership in Europe*. Chicago, IL: University of Chicago Press.

Squire, V. (2009) *The Exclusionary Politics of Asylum, Migration, Minorities, and Citizenship*. Basingstoke: Palgrave Macmillan.

Stefanski, V.M. (2009), 'Zwangsarbeit in Berlin: Erinnerungen von Polinnen und Polen im digitalen Archiv "Zwangsarbeit 1939–1945. Erinnerungen und Geschichte"', unpublished paper presented at a conference on 'Poles in Berlin', Kraków: Jagiellonian University (March).

Stefansson, A. (2004) 'Homecomings to the Future: From Diasporic Mythographies to Social Projects of Return', in F. Markowitz and A. Stefansson (eds), *Homecomings: Unsettling Paths of Return*. Lanham, MD: Lexington Books, pp. 2–20.

Stiglitz, J. (1994) *Whither Socialism*. Cambridge, MA: MIT Press.

Štiks, I. (2006) 'Nationality and Citizenship in the Former Yugoslavia: From Disintegration to the European Integration', *South East European and Black Sea Studies*, 6(4): 483–500.

Štiks, I. (2010) 'The Citizenship Conundrum in Post-Communist Europe: The Instructive Case of Croatia', *Europe-Asia Studies*, 62(10): 1621–38.

Suny, R.G. (1993) *Looking towards Ararat: Armenia in Modern History*. Bloomington, IN: Indiana University Press.

Szaruga, L. (2008–2009) 'Tor zu einer anderen Welt', *Dialog. Deutsch-Polnisches Magazin*, No. 85–86.

Ter-Minassian, T. (2007) *Erevan: La Construction d'une Capitale à l'Epoque Sovietique*. Presses Univ. de Rennes.

Tintori, G. (2011) 'The Transnational Political Practices of "Latin American Italians"', *International Migration*, 49(3): 170–88.

Tishkov, V. (1997) *Ethnicity, Nationalism and Conflict in and after the Soviet Union. The Mind Aflame*. London: Sage Publications.

Tishkov, V. (2003) 'Uvlechenie diasporoi (o politicheskikh smyslakh diasporalnogo diskursa)', *Diaspory*, 2: 160–83.

Tokarczuk, O. (2008–2009) 'Drei Gründe, warum ich Berlin mag', *Dialog. Deutsch-Polnisches Magazin*, No. 85–86.

Tölölyan, K. (1996) 'Rethinking Diaspora(s): Stateless Power in the Transnational Moment', *Diaspora*, 5(1): 3–36.

Torpey, J. (2000) *The Invention of the Passport: Surveillance, Citizenship and the State*. Cambridge: Cambridge University Press.

Toulmin, S. (1990) *Cosmopolis: The Hidden Agenda of Modernity*. New York: Free Press.

Turner, B.S. (1990) 'Outline of a Theory of Citizenship', *Sociology*, 24: 189–217.

UNHCR, R.B.f.E. (1997) 'Citizenship and Prevention of Statelessness Linked to the Disintegration of the Socialist Federal Republic of Yugoslavia', *European Series*, 3(1 – June).

Urry, J. (2000) *Sociology beyond Societies: Mobilities for the Twenty-first Century*. London: Routledge.

Varadarajan, L. (2010) *The Domestic Abroad: Diasporas in International Relations*. Oxford: Oxford University Press.

Vertovec, S. (2007) 'Super-diversity and its Implications', *Ethnic and Racial Studies*, 30(6): 1024–54.

Vertovec, S. (2009) 'Cosmopolitanism in Attitude, Practice and Competence', *MMG Working Paper*, 09–08. Online. Available at: www.mmg.mpg.de/workingspapers (accessed 12 January 2011).

Vertovec, S. and R. Cohen (eds) (2002) *Conceiving Cosmopolitanism: Theory, Context and Practice*. Oxford: Oxford University Press.

Vogt, C.E. (2007) *Nansens Kamp mot Hungersnöden i Russland 1921–23*. Oslo: Aschehoug.

Walaszek, A. (ed.) (2001) *Polska diaspora*. Kraków: Wydawnictwo Literackie.

Weinar, A. (2008) 'Diaspora as an Actor of Migration Policy', Warszawa: CMR Working Papers. Online. Available at: http://www.migracje.uw.edu.pl/publ/612/ (accessed 19 October 2011).

Werbner, P. (1999) 'Global Pathways. Working Class Cosmopolitans and the Creation of Transnational Ethnic Worlds', *Social Anthropology*, 7(1): 17–35.

Werbner, P. (2005) 'The Predicament of Diaspora and Millennial Islam. Reflections on September 11: 2001', in S. Nökel and L. Tezcan (eds), *Islam and the New Europe*. Bielefeld: Transcript, pp. 127–51.

Werbner, P. (2006) 'Understanding Vernacular Cosmopolitanism', *Anthropology News*, 47(5): 7–11.

Werbner, P. (2008a) 'Introduction: Towards a New Cosmopolitan Anthropology', in P. Werbner (ed.), *Anthropology and the New Cosmopolitanism*. Oxford: Berg, pp. 1–31.

Werbner, P. (2008b) *Anthropology and the New Cosmopolitanism*. Oxford: Berg.

Wesley, K. (2010) *Plant an Idea, Plant a Tree*, 2nd edn. Yerevan and Watertown: ATP Newsletter. Online. Available at: www.armeniatree.org/whatwedo/what.htm (accessed 27 March 2011).

Wessendorf, S. (2009) 'Roots-migrants': Transnationalism and 'Return' among Second-generation Italians in Switzerland', *Journal of Ethnic and Migration Studies*, 33(7): 1083–102.

Widdowfield, R. (2000) 'The Place of Emotions in Academic Research', *Area*, 32(2): 199–208.

Wimmer, A. and N. Glick Schiller (2003) 'Methodological Nationalism, the Social Sciences, and the Study of Migration: An Essay in Historical Epistemology', *International Migration Review*, 37(3): 576–610.

Winland, D.N. (2007) *We are Now a Nation: Croats between 'Home' and 'Homeland', Anthropological Horizons*. Toronto: University of Toronto Press.

Woodward, S.L. (1995) *Balkan Tragedy: Chaos and Dissolution after the Cold War*. Washington, DC: Brookings Institution Press.

Zeitler, K.P. (2000) *Deutschlands Rolle bei der völkerrechtlichen Anerkennung der Republik Kroatien unter besonderer Berücksichtigung des deutschen Auβenministers Genscher*. Marburg: Tectum Verlag.

Ziemer, U. (2009) 'Narratives of Translocation, Dislocation and Location: Armenian Youth Cultural Identities in Southern Russia', *Europe-Asia Studies*, 61(3): 409–33.

Ziemer, U. (2011) *Ethnic Belonging, Gender and Cultural Practices: Youth Identities in Contemporary Russia*. Stuttgart: Ibidem Verlag.

Zimmermann, W. (1996) *Origins of a Catastrophe: Yugoslavia and its Destroyers*. New York: Times Books.

Žižek, S. (1990) 'Eastern Europe's Republics of Gilead', *New Left Review*, 183: 50–62.

Zolberg, A.R. (1983) 'The Formation of New States as a Refugee-generating Process', *Annals of the American Academy of Political and Social Sciences*, 467: 24–38.

Index